Little BIG
think

Short stories to digest in the mind

Mattees van Dijk

Preface

This book contains short philosophical reflections about consciousness, time, matter and mind, the universe and God.
It is a compilation of the many blogs on my Facebook page that I wrote from 2015 to 2018 with 13000 followers mainly from India, USA and Europe.
In 2012 I published my first book "Chaos, order and consciousness". This book represents my thoughts about the fundamental role of consciousness and its relation to reality.

About the author

Mattees van Dijk was born in 1953 as the first son of a dentist. He grew up in the Westland, a small region in the Netherlands with a vast greenhouse horticulture. He studied medicine in Rotterdam's Erasmus University and started an office as a general practitioner in Maassluis. Later in his life he moved to Hoek of Holland also as a GP until his retirement in 2018. In his career he became an expert in diabetes type 2, but because of his interest in philosophy of the mind he decided to write a book called "Chaos, order and consciousness" which he published in 2012.

Special thanks

I owe special thanks to my brother Advaita Das, devoted hindu, for his advice about my book in the more spiritual sense.
I also owe many thanks to Hans van Dissel , my cousin, for his review in the translation to the proper English language. My dear friend and collegue Hubert Derksen , a reader of science fiction novels, especially Asimov, for the review of the content of this book.

Content

Chapter one

Reality isn't what we observe

Everybody knows your brain is fooling you. A good example is an optical
illusion.
But there's much more. Your brain also manages to synchronize sound and vision.
All observations in all senses are particularly synchronized in a way that they
create a simulation of reality. It's important to realize this synchronization
process leads to some delay, so what you experience is not in real time. We
know we are behind at least half a second, but some scientists have proved even
longer, even three seconds.
So the world we believe to live in is a sort of a movie created by our brain. But a
movie in which all players agree that they are in exactly the same scenario It's a
human movie, because animals each have another one. Because animals don't
have the gift of speech, they can't tell us their reality and therefore our human
reality is the only one known to us.

Plato, the famous philosopher in ancient Greece described our world as one of
looking to the shadows in a cave. Outside was reality and the cavemen were
only watching the shadows of it. Since then we call our reality the Platonic
world view.
With the scientific revolution, a couple of centuries ago, we have been able to
analyze our world with physics and mathematics. Modern science uses the method
of reduction in which one explanation prevails until it's falsified by another one.
Because of the predictability of our world of cause and effect, we call this the
deterministic world view.
If you know how cause and effect are related to each other by natural laws you
could in theory wind back the film to the beginning of time. With the invention of
the telescope we saw that natural laws are universal, because other stars behave in
the same way as ours. We are certain natural laws are, at every level from atom to
galaxies, universal.

Einstein made a grand change in our world view.
He studied gravity, time and mass of light and electrons and he launched his
famous theory of relativity. Time wasn't constant, but a relative factor in a
gravity field.

Clocks are not running the same everywhere in the universe. Space could be expressed as time. Light beams travel in three dimensions with the maximum speed in the universe, so the fourth dimension of space was time, which he called spacetime. Though Einstein was convinced that time was an illusion, he explained our experience of time by entropy. Entropy is the state of disorder the universe is in. The universe is developing from a state of low entropy to high entropy. A good example is a vase falling on a floor shattered to pieces. If you rewind the film of this event you see a vase shattered in pieces getting back in its original state of a complete vase. As we recognize this as impossible we know the arrow of time goes from cause (complete vase with low entropy) to effect (shattered vase with high entropy).

So I described two worlds now : the Platonic world (which is a simulation in your brain) and the deterministic world (which is the world of cause and effect) driven by entropy.
About a century ago a third world was discovered, even more fascinating.
The discovery was made in experiments with light. Light is a wave function, but also a particle (photon) and the problem seemed to be that they existed together at the same time (complementarity). A photon is a particle without mass travelling with the speed of light in the vacuum, but is also a quantum , a piece of energy with a spin, an energy potential and a speed somewhere in a space. Scientists discovered that one photon, or an electron could be at two places together (superposition), could change position without using time (quantum leap) or change information through the whole universe without using time (entanglement).

Furthermore, if you measure the place of a photon you will not know its speed or spin and the other way around.

Max Planck, the father of the quantum theory, suggested the universe as we knew it wasn't deterministic at all, but completely probabilistic.
Events do not happen from cause to effect, but as a function of a statistic possibility.
In a famous experiment Niels Bohr proved that the quantum state of the universe could be transformed to a deterministic one by observation or measurement of a quantum system. Our eyes are measurement devices for light, so looking to our world makes the quantum world disappear and change in to a deterministic one. (The quantum collapse). In other words, our world exists as real if watched upon.
Einstein always denied these findings, but he was wrong.

A famous thought experiment is Schrödinger's cat (called after the great physicist Erwin Schrodinger), where a non-observed cat lies in a closed box with a device that works on decay of atoms (which is a quantum probability). This device, when activated by decay, would smash a capsule with gaseous poison and thus kill the cat. In the quantum world there's superposition, so both probabilities: a dead cat or a living one would be happening at the same time in the closed box.

If you would open the box you would observe either a dead cat or a living one, not both (quantum collapse). This view is called the many world philosophy. In theory every observation of a quantum state would lead to one universe with a dead cat and one with a living one. Those parallel worlds do not exist as we know, because we only observe ours, so what's going on here?

There are a lot of problems with this interpretation of quantum mechanics. Universes don't split up with every observation of a superposed quantum event. So, we have to reject this view.

Many physicists deny that consciousness is the trigger for the quantum collapse. They argue that it's the measurement device that does. But the measurement device is made by sentient beings and the interpretation of the measurement is too.

Moreover, the retina of the eye is also a quantum measurement device.

Philosophers do also think that consciousness is collapsing the wave-function of the quantum state, but they're also divided in their opinion. How can consciousness collapse the quantum state? It would mean that reality is only there if we watch it.

So if we don't look at the moon, it's not there, which is nonsense of course. Humans have only been a minute moment in the universe to observe it, so if there was no conscious observer, wouldn't there be a universe older than humanity? Of course not, we see very old stars and know the universe has a lifespan of 13 billion years. If you argue that conscious observation is the trigger for the collapse then you must accept there's an observer everywhere in the universe. Suggesting that consciousness is everywhere. Consciousness must be a fundamental property of the universe. I call this holo-consciousness.

In my book "Chaos, order and consciousness" I have called the quantum world the super reality state of the universe. We can't experience this particular universal state because our consciousness is collapsing it to the normal universe we know. In the super reality there's no time, no space and many probable worlds can exist together (parallel worlds), everything that might happen, happens indeed.

Because this world wouldn't be inhabitable for conscious beings, time, space, cause and effect, entropy, the arrow of time is introduced by holo-consciousness as a flow of experience. We experience this because of our brains that are able to link somehow with holo-consciousness.

Though we have the illusion that coincidence is driving the universe, this is not true, it even doesn't exist at all. To get a meaningful universe you need observers with consciousness. This means complexity, ingenuity, evolution and the ignition of consciousness in individual sentient beings.

There's no room for coincidence in this delicate mix of chaos, order and consciousness, because it would mess up things completely.

Everything is directed by consciousness, consciousness is all, it is all there is.

Chapter two

The nature of consciousness is natural, not mystical

John Searle is a famous philosopher of the mind.
Scientists think , you can't study consciousness, but Searle and many others think you can, be it on a different level.
This wrong opinion is attacked by him during a speech at the CERN, which is the particle accelerator in Geneva, the Walhalla of science.
Every year there are lectures about different subjects by the most talented and gifted minds of the world, usually about exact science.
Searle was received with skepticism , because most of the audience have a natural science education and no interest in woolly talk about something mystical as consciousness. But Searle gets their attention quickly when he makes clear that consciousness is a natural phenomenon, just like energy or mass and is not woolly at all, you just can't measure it with units or mathematics.
He made quite an impression on the highly intelligent audience.
He continued with categorizing consciousness in three possible theories:

1. It doesn't exist, it's an illusion
2. It's like a computer program, just information processing
3. It's the most fundamental property of the universe

If consciousness did not exist it would be the same as saying that don't exist at all. Consciousness is always there, the minute that you awake, when you dream, you can't switch it off like that. It's just there. We're not zombies, nor machines. So, this theory is firmly rejected.
Consciousness is not a computer program either, because computers are inanimate things, pull out the plug and nothing remains, it's just a dead thing, a calculator. The living brain is working on a completely other principle and it is even doubtful that the brain is even really having experiences itself. It's unlikely that the activity of neurons alone causes immaterial phenomena like consciousness More likely is that consciousness is a universal phenomenon, belonging to nature and the cosmos, just like energy, time and entropy.

The cosmos can only exist by the grace of quantum fields and consciousness observing them and collapsing them to matter. No consciousness, no cosmos.

Searle, having an audience from exact science harvested much acclaim with his view on consciousness. Searle was surprised that we could build complex particle accelerators, send robots to other planets, develop artificial intelligence, but that we're incapable to analyze consciousness, a phenomenon which we experience all our life and is as common and as natural as anything else. If we could manage to solve this problem, we might find the basis of all science.

Chapter three

Holistic science and pantheism

Officially there's no such thing as holistic science. Science is reductionistic in nature.

Holistic science does claim that a studied object isn't the sum of its parts, but a whole thing with a meaning. For example, our body isn't the sum of trillions and trillions of atoms, but a complete person. A painting isn't the sum of pigments but is a piece of art.

Our brain, for example, is a sort of cauliflower-like organ with billions of neurons and trillions of synapses, which is of great importance to us. We don't see ourselves as a brain and indeed we aren't. Many scientists think that consciousness is situated between the bones of our skull, but there's no scientific proof for that at all. The problem with consciousness is that there's no way to measure it, nor any mathematical or physical approach will do so. We don't have consciousness-measurement devices.

Though some neuroscientists really think they see consciousness on brain scans, which is quite naïve of course. In fact we encounter the border of science. The only approach for this is a philosophical one or developing the best possible scientific theory for consciousness. Welcome to holistic science.

We are sure about some facts. No brain: no conscious experience. No cosmos: no life. No life: no brain. Nobody ever got back to consciousness having died before.

We have reported near-death experiences and know that the brain sometimes can be very active ten minutes after a person died. But that's about it. We also know that your brain can be comatose but still conscious (locked in syndrome). You can lose 90% of brain tissue, but still function normally. The brain can be completely functional but you can be in an unfunctional state like a psychotic depression, which is as good as experiencing nothing. So, there's a link between consciousness and the brain, but it's not an exclusive relation. Holistic science might postulate a hypothesis and work that out.

What else could be the subject of holistic science?
Some scientist suggests time travel might be possible theoretically , for example through worm holes (a giant curvature of time space sucking everything in on one side and throwing everything out on the other side, displacing your space ship to another place and time).Still we must consider the fact that nobody ever

came out of the future or past to visit us, so holistically this is a forbidden event in the universe.

Reasoning through logic we must conclude too that it would be impossible to visit your own funeral via time travel in the future, because you would be dead and alive at the same time. This is called the time paradox, but is it really a paradox? It's just the outcome of logic.

Another important subject is evolution. The scientific research on this matter is so vast that it's a scientific fact that evolution took place and creationism is rejected. But still there's no proof that it all took place by accident, instead we discover patterns in nature that are similar to natural laws which claim forbidden events and compulsory events. Life has to procreate (compulsory), has to develop from simple to complex (compulsory), and can't live eternally (forbidden). So, nature itself certainly has some intelligence and self-organization. This doesn't sound counter intuitive for us, because we tenderly speak about "Mother nature". But it also suggests that we feel that nature has a mind of its own.

Our planet certainly is believed to be a special place in the universe where everything was just right to evolve life forms, even intelligent life forms. The vast universe is a dead place and it's unlikely that there are many places were life can exist. Still the vastness of the universe, with its trillions of planets makes it very unlikely that we're all alone and despite that teaming with life. The general scientific view does suggest that our existence is a "freak accident".

"Why", is a holistic question! We know the universe is made of different forces of gravity, electro-magnetism, weak nuclear force and strong nuclear force. These forces have constants that are so tuned to each other that they can form stars, planets and even life. So, is it really a freak accident? Or a mathematical fine- tuned system? Many think the universe is anthropic, which means made for life, also called biocentrism. So, it might be that on millions of planets life has emerged and on thousands of them even intelligent life at a certain time. It suggests that the universe has some "do's" and "don'ts" too. Forbidden and compulsory events. The universe looks self-organizing too.

If we look at our planet again then we see that nature makes ingenious designs.

How about the eye? The chlorophyll molecule? Plants making oxygen for animals and animals making dioxide for plants? The chameleon's camouflage strategy? Or the walking branch? The simple spider making a perfect mathematical web? Etcetera. We humans copy nature in our technology, many examples can be given, but I only mention the bird and the airplane. Is nature rudderless, meaningless and accidental? How can you maintain that opinion, looking and studying nature? Of course, Darwin gave us a clue and many researches and the discovery of DNA gave us much information about how evolution works, but the outcome is just too awesome to suspect rudderless coincidence. The holistic approach wants to define the forbidden and compulsory events in evolution too, accepting that nature is self- organizing and has some kind of mind too.

In science we use axioms. These are statements that you can't prove, but all agree upon as being true. For example, the shortest distance between two points is a straight line.

We all agree on, but we can't prove it! If you bend space with gravity this line might not be as straight, so what is even the value of this axiom? Why not accept then that the universe, nature, has a mind of itself as an axiom? Why not accept the universe is there for us, for sentient beings that ask these questions?

Why not?

Of course you would say this is all speculation, which is true. But has speculation not brought us great science? Holistic science meets science, philosophy, theology and even religion here.

Pantheism is the belief that God and the universe are the same. This God has nothing to do with humanity in particular. You can't pray to Him for support, comfort or some favor. It doesn't have messaging prophets, didn't write holy books. It even is not a person or a mind that thinks human. It's just a necessity for the universe to exist. It's the supportive processor of all information available in the universe This God doesn't break any natural laws. It can't let miracles happen. It doesn't appear in human form.

Maybe the universe was made coincidental, but I'm not convinced at all. Was it a transition from the timeless quantum universe to the time synchronized universe by an initial quantum collapse? Who was the primal observer? The first measurer of the quantum universe? God? Holo-consciousness? I bet on the last one.

Chapter four

Holistic science versus reductionistic science

The reductionistic method (R) is the principle that science should be limited to studying objective facts and the relation between cause and effect, but does not allow a subjective opinion about it or any meaning.

The holistic method (H) does acknowledge scientific facts, but has an opinion about its place in the whole picture. To show the differences, this table might help:

Holistic:(H)

Big bang
The start of the universe, probably by a first collapse of the quantum state by the first conscious observer.

Coincidence
Doesn't exist, is an illusion

God
May exist

Free will
Existent, but limited by subconscious brain processing

Consciousness
Fundamental phenomenon of the universe

Matter

Information about a volume of space-time

Time

Illusion caused by consciousness. Past, present and future all exist, but consciousness makes it flow in a specific order.

Reality

Transcendent, being the content of consciousness

The brain

Information system being linked to holo-consciousness as a "client".

Universe

Space-time system in a state of consciousness, exactly right to shelter life and local consciousness systems (sentient beings).

Evolution

Purposeful process of a multi-population of living beings that develop from simple to complex.

Meaning of life

Maybe an evolutionary goal, but for a normal individual the most important goal is to get love and attention, stay healthy and be happy.

Reductionistic (R)

Big bang
A coincidental quantum fluctuation leading to a material world

Coincidence
Existent

God
Non-existent

Free will
Non-existent

Consciousness
Illusion fabricated by the brain

Reality
Physical existence outside the mind body system of sentient beings

Brain
Living organ that controls body and mind; in fact, you are your brain.

Matter
Elementary particles

Time
Flow of events in a specific order on the basis of causality.

Universe
One of the many universes in which an observer can develop (multiverse theory)

Evolution
The organisation of living beings by natural selection alone in which there is no direction at all, no special meaning, only adaptation.

Meaning of life-
Useless and senseless

Of course this table isn't completely free from subjectivity.
In the first place the holistic version is my personal opinion and can be refuted by others with good arguments.
Secondly science is not so sure about many things I mentioned here, so most of the answers should be: *"We actually don't know"*.

So, the holistic method is very subjective and leans on philosophy, the time spirit and the opinion of leading thinkers. Also, the holistic method speculates on argumentations, thought experiments and the most leading opinion. An advantage of the holistic method is that it integrates great scientific theories and is "getting the grand picture".

The deterministic method is very effective, but fails in explaining the "big questions". The enormous success of the reductionistic method is that it is shared by many and is falsifiable, testable by experiments and mostly tangible and concrete.
A disadvantage of the reductionistic method is that it's too much specialized and segregates the different sciences, because of which no "big picture" appears.

Of course methods overlap and are not opposite to each other. It's not a "black and white" contrast.
Though general scientific thinking should be very objective, we are humans and by nature irrational and subjective. The egos of leading scientists are often a disturbing factor in objectivity, but group pressure can also be very oppressive. For example, atheism can be a factor for a successful scientific career in some disciplines, which is not objective. Funding and nepotism can also corrupt the scientific world.

The philosophical approach of holistic science is elegant, though we have to accept premises to use this method. Personally, I have the feeling science is crashing under its own success and is not working out the big questions due to lack of courage.

Chapter five

Time isn't real

In his time Einstein had difficulty explaining time to his colleagues. With his brilliant mind and as a master of abstract thinking he hardly could explain that time is an illusion. We experience time as a flow of events in a specific order from cause to effect and this feels completely natural to us. How could this possibly be an illusion? He explained that as time doesn't exist as a real property of the universe everything happens at once. How absurd and weird this idea was! Past, present and future would exist together without any passing of time, which is incomprehensible. He proved that clocks didn't work at the same rate in the universe and time was just a function of gravity and the speed of light. If you could travel together with a photon you wouldn't experience any time at all even if that photon started traveling at the mere beginning of the universe, some 13 billion years ago. Moreover, time would slow down or speed up with the relative speed of an object and the strength of the force of gravity. On the edge of a black hole (event horizon) time also stands still because of the immense force of gravity. In other words, the slower you go and the less gravity forces works on you the more time you experience. In our world, of course, we go very very slowly (in comparison with light speed) and don't have extreme forces of gravity and that's the reason we experience time flowing abundantly.
Einstein's discovery has never been falsified for 102 years now.

In my book I also emphasize that just like other phenomena (energy, matter, space) time is just information in "the database" we call the universe.
Information is the basis of the universe.
Atoms, quarks, photons, electrons, don't exist as particles, but are just bits of information belonging to a bit of space-time. And space-time itself is also just information. Particles are not tiny ping pong balls made of undividable matter. Information is as such also undefinable, abstract and non-local, we can't localize it at some point in the universe, it is just everywhere and at every point. In our world we find information everywhere indeed, except the information of particles we find information in DNA, computers, books, but most importantly in our own brain.
The information processing what we *experience* is called consciousness. Our brain is the information container of conscious information changing on a constructed time scale. The information in our consciousness we call qualia. Qualia are non- describable bits of information in our mind, like the color yellow, a tingling feeling in the finger or a low sound. Our brain acts as an interface between the "world out there" and our consciousness.
There's no scientific proof that consciousness is made by the brain, nor is there any proof that consciousness comes from outside the brain.

Both views have their own arguments. It seems very unlikely that physical processes like firing neurons alone, cause consciousness, which is in fact an immaterial phenomenon. On the other hand, it also seems very unlikely that consciousness is present in living beings without a central nervous system. My opinion is in between both possibilities. I'm convinced conscious is non-local, but also that the brain plays a central role in making it local and usable for a person as experiencing his body and his place in the world consciously. If consciousness is non- local that means it's universal. I call the universal consciousness: holo-consciousness. (holos = whole, *Greek*). If holo-consciousness is everywhere then the universe must be the container of it, because the universe is also everywhere. Just as our local brain supports our local consciousness, the universe is fundamental for holo-consciousness. The universe is built of unconscious stars and we don't see a huge brain out there in deep space, so consciousness is not measurable nor observable by scientific methods for us. Now, non-locality does also mean eternality. Because time and space just belong to the same set of dimensions.

The mystery is why do we experience time because of our brain function, while the holo- consciousness is eternal and non-local? Now how could the brain accomplish that?

I can only give an example of how information and time are linked.

Suppose you have made a trip to Italy and you made slides. Now "all" the information of your trip is in that box with slides when you take them out of the cupboard. The right order from the trip from Milano to Napoli is only presented after you put the pictures into the right order by your own acts. So, some "process" is making the time flow from cause to effect, from Milano to Napoli. It's your brain that's doing the processing and your consciousness is consuming it. Time as the illusion of the right sequence.

If time is an illusion of the sequence from cause to effect made by some process in our brains or some process in the universe what's the processing rule and how does the system know what's the cause and what's the effect? Newton explained that to us by giving mathematical formulas for natural laws like gravity, force, mass and energy.

Thanks to his physics we can launch rockets to the moon, predict when Jupiter is in opposition and where that tennis ball will hit the gravel at Wimbledon, with great precision. This form of physics is called classical physics in which we all recognize the world which we are so familiar with. In fact, we could exactly predict how the universe develops from cause to effect if we had all the information of it. (Which we don't of course) We have a clockwork universe which determines its faith at the moment of birth (the big bang).

Though there's much controversy about the role of consciousness in quantum physics this doesn't mean you can't rule out that consciousness is fundamental in reality. On the contrary! A strong argument is that quanta are the smallest packets of energy and information existing. Because the universe is made of quanta it seems reasonable that the quantum collapse happens all the time because of observation by holo-consciousness. Quantum physics shook up our sense of reality and showed time is probably a derived product of the permanent universal quantum collapse.

Though we can't visualize the quantum world directly, there's no scientific doubt whatsoever that the quantum world is real. So, our universe is a quantum universe consisting not of particles but of quantum fields. This is a universal energy grid with probabilities telling us that "something like a particle" might be there. Just by looking at it, it appears into our classical Newtonian reality. So, the table you see in your house is the sum of probabilities of particles forming a table! Now let's imagine ourselves that we could shrink ourselves to a billionth of a billionth of a billionth of a billionth of a billionth millimeter.

What would we see? The answer is: empty space. We are in Planck space, an extremely tiny part of the universe where time nor space has any meaning. There will be no event in that part of space as long as the universe exists. Of course, the same way of reasoning you can apply to time. There is a fragment of a second in which no time will ever flow (Planck second). If the universe is the sum of all these Planck spaces and Planck seconds you can conclude nothing happens and there's no space anywhere, which is absurd of course. The probability of one Planck space for an event to happen is practically zero, but not absolutely zero and there is a mind-blowing amount of Planck spaces in the universe. The probability that the universe exits is of course 1, which is the ultimate sum of those innumerable Planck spaces.

Now it's clear what the nature of time is. It's the probability of something to happen in a specific order in all Planck spaces existing in the universe. Time is invented by conscious reasoning in our brain. It's not really there, but just by observation of quantum reality it will pop out as a flow of events in the right order from cause to effect in space-time. Picked up in pieces , like the Italian slides and put together in the right sequence by consciousness.

Of course, our brain is not really doing that, it would be too much to knot quantum events, which are fundamentally timeless, together, let alone measure each probability of the event.

Some more powerful conscious phenomenon can do that of course : holo-consciousness.

But there are rules for it and they are not only Newtonian. To understand these rules, we must know the second law of thermodynamics from Boltzmann.

Boltzmann discovered "the arrow of time" (the cause to effect relation). Events that are directly related to each other by cause and effect will always be placed in the specific sequence from cause followed by effect. This is because the universe goes from low to high entropy. Now entropy is quite an abstract concept.

Entropy is in fact the amount of chaos in the universe. So, going from low to high entropy means that an orderly universe will change into a chaotic one. In daily life we can see this happening when we see a cup of tea fall to the ground shattering into pieces and splashing all the tea around. The cup of tea (order/ low entropy) changed into shatters and splashes of tea on the ground (chaos / high entropy). It's unthinkable that these splashes of tea and those shatters from the cup will spontaneously get together perfectly again in the cup of tea. The first orderly state (cup of tea) was in an equilibrial state of low entropy (order) and the smashed version in an equilibrial state of high entropy (chaos). In the cosmos we see this happen on a large scale. Stars explode as supernovae and send their debris into deep space; which eons later will form new stars again.

Now in the example from the supernovae we see that almost from every "effect" we can expect a new one. So, the "effect" of an exploding star caused by the implosion under its own gravity is the next "cause" of new star forming. So, effects turn to cause and so on until a new equilibrium is reached. The clue is that an orderly state has a lot of spontaneous potential cause-to-effect cycles in it and a chaotic state none. (Imagine the cup of tea).

What this means is that the universe is gradually running out of cause-to-effect cycles, because high entropy will take over in the universe as the new equilibrium. At the end nothing happens anymore. Black holes stop radiating, nothing moves relatively to each other, absolute zero energy is reached, quarks disintegrate into nothingness and the universe dies a heath death Most importantly time will stop. So, the conclusion is also that time and entropy have a constant relation to each other. If entropy is changing from low to high, there's time. No change of entropy, no event, no time. But time can only be constructed by consciousness, because it's the illusion of consciousness. So, entropy and consciousness are also related.

If nothing happens in the cosmos, consciousness cannot experience time. If there's no time, there can be no cosmos. So, if there's no consciousness there's no working cosmos either by lack of conscious observation. This would mean that consciousness is the central element in all existence.

Now Einstein rejected quantum physics to the end of his life. Not because the mathematics were wrong, but because quantum physics didn't match with the theory of relativity. It's a shame that Einstein, certainly the greatest scientist that ever lived, didn't try to match the two together. It was his philosophical opinion that consciousness might be the missing link between the space-time reality and the quantum reality. (or the quantum universe and the classical universe) Niels Bohr, his opponent and father of quantum mechanics, had many interesting and witty debates with Einstein about this issue. Also, Max Planck believed that consciousness translated the quantum universe to the space- time universe (quantum collapse). Both masterminds raved with Buddhism which believes holo-consciousness is the only real phenomenon and the universe is only the derivate from it.

Because this is not science anymore, until now, scientists haven't solved this major problem.

Chapter six

The soul doesn't exist

Admit that almost every day you exist, you probably wonder some seconds of the day who you are and where you're heading to. Our ego is our most valued possession which we experience as unique, equipped with a free will and being the center of everything.

But face it, you know better, because you're sharing this planet with billions of other humans who are also the center of everything. A planet, by the way, that's just a tiny blue spot in a vast universe. You certainly know you're almost nothing in the grand scheme of everything. You live your life in the blink of an eye being in a cosmos billions of years old, aware that you're mortal and vulnerable. You're experiencing yourself as the center of your world because your eyes and your brain are carefully placed at the top of your body and your head feels as the place where you actually are. So, the brain is making you localized in a head. This brain, our processing machine that controls your body and receives information of your environment is also a vulnerable organ, just as vulnerable as your liver, kidney, heart or lungs. Our sanctuary can be inhabited by demons when you get a psychiatric disease like schizophrenia, anxiety disorder or a depression. A little puny embolus can destroy an essential part of your brain and making you cripple or mute. And what about a brain that shrinks like a grape into a currant which slowly kills your personality, social life and dignity in Alzheimer's disease?

Which holy fire is burning inside that organ that makes you the centre of everything?

Is it the soul, the metaphysical "self" that makes you exist?

Is the soul taking possession of your brain and leaving it when you die?

It was the philosopher Descartes in the time of rationalism, who delved into this problem and tried to analyze it and founded the philosophy of the mind. Descartes assumed that the mind and the body were separate things (dualism) and he thought the pineapple gland was homing "the self". This small structure hangs down the base of the brain wedged between the two optical nerves.

He fantasized about a homunculus (little man) living there and controlling the brain and body. If so, others opposed, who is controlling the body and brain of the homunculus, another homunculus in his pineapple gland, etcetera? He also tried to weigh the soul by weighing a dying person before and directly after death. If the soul left the body the body should be lighter. Of course, he failed to prove that.

It was also Descartes who spoke the famous slogan in Latin: "cogito ergo sum". "I think and therefore I am".

Now of course it's easy to say Descartes had primitive thoughts about self-awareness, but with our present knowledge of brain function that's quite a cheap thing to say. We know the pineapple gland (hypophysis) is a very active gland that makes master hormones for many other glands in the body mainly the thyroid gland, the kidney and the gonadal glands. Still his thoughts about thinking and being aware are very important. We know now that the cortex and especially the neocortex is important to feel self-aware. It's a process and not a homunculus. The self is a construct of our waking or dreaming brain. As long as you live this self-awareness is present, it's just impossible not to have a self (except in catatonia, a psychiatric condition).
But are we our brain? Who am I, an organ?
Is the reader of this text just an organ? It's my brain that *makes me think*, but it's *me* who does the thinking in free will, isn't it? Is my brain my servant or is my "self" the illusion that is made by the brain?

To approach this question we first have to know what our brain actually does. Primarily our brain translates the reality "out there" into a concept that is universal (we think) for all humans. So, we experience the color yellow and name it "yellow" in our own language. But actually, we don't know if there is an individual variety of the color yellow for you, for me and all the other humans. We don't know because we name it all the same and can't have a look in someone else's brain.
Most philosophers state that the physical reality is not known, because our brain translates it into our human reality. Most physicists are convinced the best concept of reality is described by mathematics, but they admit that mathematics is also a human invention. They are right that square circles don't exist in nature, but also admit that it's just a matter of definition.
So even if our brain translates physical reality to navigate in it, it's just a version that's the most practical one for us to live in and to survive in.
Secondly the brain is adding a quality to the experience as such. For music is only a set of tones in a specific order and volume, in fact vibrating air and eardrums, but it has a quality for us and is moving and touching us.
The same applies to reading a book, you fantasize a situation "in your head" which fits to the symbols you read in the book.
Thirdly the brain makes you self-aware, there's no question about it that it's you listening to the music, seeing the yellow wall, while reading a book.

We still don't understand how our brain does the trick, but we have some clue. Most of the brain work is silent and goes unnoticed, automatic and completely unconscious or subconscious. This might be 99,9999 % of all brain activity. The state of consciousness is reached after more than half a second or even some seconds as a report of experienced reality. The main task of the brain is called the process of "exformation", not just processing information. Exformation is in fact the "throwing away" of information that's not important enough to reach the state of our consciousness. So, consciousness is a state in which a summary is given as a report afterwards to you. If so, is my subconscious brain doing my work and I'm just being an observer, like a man in a balloon? A guest in a brain? Hard-line neuroscientists might think that, but of course it's a false conclusion.

The misunderstanding of time (see previous chapter) and our own personal understanding of the world (widely dependent on brain function) are making it very unlikely that we are just an honored guest in our own brain.

Consciousness is a continuum, a flow of experiences, not belonging to the subconscious processes of the brain. In fact, those subconscious processes are conditional for the state of consciousness, but it's the conscious state itself that's making the decisions. So simply explained conscious state A for person X leads to decision B and the same state A for person Y can lead to decision C.

This is what we call free will. It's not a random process but caused by consciousness prepared by our brain.

Having said that it's also quite clear that self-awareness can't be explained without an active brain. Self-awareness is also very rare in nature. Many primates are self-aware, elephants, dolphins, whales and maybe horses and pets. Substantial evidence for higher primates, but not for many other animals. We don't know what the evolutionary advantage of self-awareness is. For a human the fact that you know you will die at a certain moment can be a cruel side effect of self-awareness.

The most important experiments come from neurosurgery. Patients with epileptic fits were cured by a callosotomia, which is the splitting of the right brain from the left brain. These interventions didn't lead to a split personality in a right "me" and a left "me". But certain functions were split up, so lacking coordination between the left and right hemisphere lead to misunderstanding things in daily life, but the person kept feeling as one "Me".

Siamese twins with the two brains merged together, surviving birth, were also two persons and not one with two bodies.

Certain psychiatric conditions can split up a person's mind into different personalities especially in severely traumatized patients, but investigation showed these states are theatrical and not real "me" states (a psychotic state) being mostly subconsciously driven.

Psychiatry gives us much support about how our ego will resists certain brain states, like obsessive compulsive disorder, anxiety disorder, borderline, schizophrenia, depression and manic states. The original "persona" is terrified by what the sick brain does to him/her! This presupposes that the ego is more than a state of consciousness caused by the brain alone. Being your own self again after a severe psychiatric episode you will make the utmost effort to keep control over your own brain.

So, we have to define the soul, the mind, consciousness and the self.
The mind is the conscious state of the self.
The self is a lifelong concept of belonging to a localized body and mind.

Consciousness is a brain state in which experiences flow through our mind. But what about the soul? The soul is mainly a spiritual concept. Our soul experiences the after-live or reincarnation and is judged by God to be expelled from heaven or admitted to it. This typical dualistic reasoning doesn't get much support anymore since neuroscience made so much progress.
Another definition of the soul is the "deeper self ", which is typically and uniquely "you". It's clear now this deeper self is depending on brain function mostly.
No brain, no self. The soul doesn't exist as a real phenomenon, it's the holistic sum of many processes of evolution, biology, culture, history, social events, genetic events, sexual events, physical conditions. The soul is a spiritual entity, consciousness is a natural phenomenon, the mind is an abstract concept, the self is an illusion. Because your soul isn't caused by your brain activity, but by many cosmic events as mentioned above, in fact your soul is just a cosmic event too. Just as a configuration of all previous cosmic events from the big bang onwards. But in the total holistic sum it's all universal consciousness acting locally in space and time.

Chapter seven

Consciousness: reality or illusion?

The discussion is going on for centuries or even thousands of years: *"Is consciousness reality or illusion?*

In the philosophy of mind two views are recognized:

- The illusionists
- The realists

The illusionists.

The illusionists are convinced that the brain makes your consciousness and your mind is controlled by it, in fact, you are your brain.
This opinion is shared by many scientists , neuroscientists, biologists, evolutionists, some doctors, physicists and some philosophers.
In my country , the Dutch professor Dick Swaab, neuroscientist, is a typical example of a scientist who claims this view, with his best-selling book: "You are your brain".
In the philosophical world Daniel Denett, philosopher of mind, is an expert who supports this view. Actually, Denett goes a little bit further by stating that consciousness isn't existing at all, it's just the activity of brain networks. In fact, he sees consciousness as an epiphenomenon, a side effect of brain function.
It shouldn't be there at all - it has no evolutionary function.

Those opinions are certainly based on the fact that we, humans, claim our cosmic importance too much, we should be more modest as an unimportant freak accident being in a hostile cosmos. We, humans, shouldn't be here it all. It's just a cosmic mistake that we're here and therefore we must be all alone in a stone-dead universe.
This group claims to have strong proof of their view, because they are supported by the findings that neurological damage of the brain causes corresponding changes in the functioning of consciousness. Therefore, consciousness and brain function are the same.
So, for example, if you get a stroke, a part of your consciousness (the spatial experience of half your body) is gone. In paraplegia your legs don't belong to your consciousness anymore.

Disturbance of brain metabolism causes coma. Neurotransmitter imbalance can cause depression or schizophrenia.

Moreover, with fMRI scans you can visualize brain activity that corresponds perfectly with the corresponding state of consciousness, like sight. The route of visual stimuli, for example, is well studied in the brain and you can analyze quite exactly the processes that generate a visual experience.

So consciousness is a process inside your brain, between the bones of your skull and experiences are constructed by brain activity alone. Of course, this is a strong argument, but it can be challenged.

The strong point is also it's weakness, because they can't explain how physical processes as the firing of neurons can cause something immaterial as consciousness. We canmeasure the bloodstream through certain brain fields with fMRI scans, but that's just physical activity, not consciousness. We simply can't measure consciousness with that technique. Another point is that experiences are depending on the outside world. It's what happens outside the skull what makes consciousness alive and awake. Not just firing neurons in a dark skull.

The realists

Their view is completely different because they reject the assumption made by the illusionists that consciousness is made by the brain.

Instead they suppose that consciousness is natural, universal and belonging to physics, especially quantum mechanics.

This theory is supported by some neuroscientists, some physicists, many philosophers, doctors and mathematicians.

Famous opinion leaders to mention are sir Roger Penrose (mathematician; UK), David Hameroff, (anesthesiologist; USA), Deepak Chopra, (neuroscientist: USA), David Chambers, (philosopher of mind; AUS).

The main view is that conscious observation makes the quantum state collapse to the classical state of space time. What we observe exists as real, what we don't observe remains in a quantum state. I refer to chapter one concerning the explanation of the quantum collapse.

Of course it would be absurd that not observing something like the moon would make it disappear if we all don't look at it one night. The moon and many stars are gigantic pieces of matter and can't disappear because we don't observe them. In quantum mechanics vast objects stay in their classical state because of coherence between the numerous quanta making up such vast objects. This also accounts for tables, chairs, small objects, microbes, viruses etc. It's not easy to understand that consciousness has still transformed those material objects from the quantum state to the classical state. Consciousness locally "carried" by brains of humans and animals is quite rare in the cosmos and also very temporary while the universe is enormous and ancient. How could they possibly be important for the state of the universe? Therefore, the realists suppose the universe should be in a conscious state itself to maintain its classical state. The view of the illusionists is explained as a fallacy, because they reason with the completely wrong presumption that brain activity itself causes consciousness.

The classical universe means that the cosmos is local and can only be travelled through with the use of time. Quantum universe means that it's non-local and can't be localized somewhere nor travelled through because space and time are non-existent.
Furthermore both versions of the universe are both ultimately built of information bits alone.
Those information bits in a classical universe contain space and time, but not in the quantum universe which is the lowest level of information in the universe.
We can compare this with a hard disc full of usable information versus a zip file in which the same information is hidden in non-unfolded dimensions.

So this group of thinkers claims that consciousness is universal and non-local and functioning as a translator to "unzip" the quantum universe into the daily universe we recognize it. In my book I call this universal consciousness: holo-consciousness.
Now the brain uses this holo-consciousness system just as the body uses water, oxygen, glucose, proteins and fats to function as a system.
The brain is like a localiser in the universal non-local network of holo-consciousness and translates quantum information to "brain language", which is a neurobiological language with electrochemical pulses. It's not the brain that thinks or feels, it's consciousness that does that, using the brain as a biological computer. So, for the realists consciousness is fundamental and creating all reality, without it there wouldn't even be a universe. The strong argument is that in this view the "hard problem" of consciousness is solved. The hard problem of consciousness is how material processes like firing neurons can make an immaterial phenomenon as consciousness. Furthermore, the quantum collapse is also a scientific fact which makes the theory more probable.

The strong point is indeed again the weak point, because how can an immaterial phenomenon as consciousness create a material reality? We can experience consciousness by translating it in "brain language" but not measure it nor know the nature of it and that's also a problem. Another problem is that this theory also reintroduces dualism to a certain extend. Not a personal soul but a universal soul inhabits our brain and unlatches when the body-brain system dies.
These realists defend their theory with the view that the universe might not be material at all, but is just a collection of information. This somewhat slippery theory could easily be conceived as an esoteric one, leading to non- scientific points of view as are common in religions like Buddhism.
Many scientists don't want to get involved with a theory that could have the slightest odor of anti-science.

As a realist I defend the theory and in my next chapter I will explain that matter doesn't exist either indeed. We completely live in a mental universe.

Chapter eight

Matter doesn't matter

The "big bang" is a mystery, just because it's not understandable how from nothing a complete universe can be created. The theory tells us that from an infinite small point, called a singularity a giant explosion, created the universe. This happened incredibly fast and space was formed much faster than light speed (inflation theory) and time started ticking. Because of the law of conservation of energy all the energy being in a very small space is still the same now, being in a very large space.

The theory also claims that besides matter also antimatter existed, but was annihilated by matter. Antimatter is the same as matter, but only with opposite charge. Antimatter is still there but is exotic. The theory of super symmetry predicts that on every particle an antiparticle exists.

Now what are particles exactly? Rutherford, a physicist from New Zealand, considered an electron as a tiny ball swirling around a large ball, the nucleus of an atom. Nuclei are built up as protons and neutrons together. Protons have a positive charge and neutrons no charge at all. When an atom has two protons (helium) then two electrons are needed to give the atom a neutral charge. Chemical bindings are created by atoms who exchange electrons with other atoms to maintain a neutral charge together. A set of atoms are molecules. So, water (H_2O) is a molecule consisting of two hydrogen atoms and one oxygen atom.

This model worked perfectly well until quantum physics came around. Electrons were not small balls circling around a nucleus, but a probability that it's somewhere over there expressed in quantum waves.

Furthermore, those probability waves tried to maintain the lowest energy level possible in an atom. These energy levels are expressed in electronvolts, so particles became energy levels.

Then, there's string theory, small strings that have their own unique form and movement, but have enrolled in them nine other (unknown) dimensions.

So these strings and these probability waves don't consist of tiny balls anymore whatsoever! Matter is defined as an energy level, spin, speed and they are all obeying quantum laws.

In fact, though matter exists as a probability of some specific property of a small part of space it's completely abstract. All the mathematical models fit perfectly, but we can never point to an electron or a photon, because we aren't sure they are there!

In fact, we can't visualize matter at all, except in large particle accelerators as charges in a measurement device. So, matter is energy and we don't know what energy actually is. We only know what energy does. So, is matter an illusion? No, of course not, it exists and we can compute it and experience matter consciously, when a stone hits our head. But is matter real?

I doubt if it's real, because it's just a description of how the universe is filled up with information. How, for example, does an electron interact with a photon? Is information real though? No, information is completely virtual. So, we must conclude that what we consider as matter is just a virtual database of information we call "the universe".

The truth is that the universe is completely empty and virtual. An immaterial universe.

So, our reality is virtual, though consciousness makes it possible to have experiences in it.

If so, consciousness is the fundamental phenomenon which makes the universe exist at all. No conscious observation, no universe, no reality.

To understand that matter is virtual we must make a trip to a black hole.

A black hole is born when a heavy large star is burnt up, explodes as a supernova and then shrinks under its own weight to a small celestial body with a mind-bending density. Consider a complete planet earth compressed to the size of a ping pong ball.

The gravity field of such a monster is so extreme that even light particles, having no mass can't escape from the black hole. This confirms Einstein's theory that gravity is just the bending of space time. In the case of a black hole you must realize that the bending of space time is so strong that it forms a deep hole in the space-time grid and as a whirlpool in your sink is sucking everything inside.

Standing next to a black hole (theoretically), we wouldn't see things disappear in the black hole, because the light rays wouldn't reach our eyes. It was the late Stephen Hawkings who true studied black holes quite accurately.

If you were standing there with a flashlight then the light ray would travel in a circle around the black hole and shine on your back! Hawking discovered that light does circle over there and is amplifying itself by eternally travelling in circles. According to the theory of relativity the force of gravity is so strong that all the clocks would come to a standstill, e.g. time would stop.

The wafer-thin zone at the edge of a black hole is called the event horizon. This is the point where time stops and, in fact all experiences stop, so the "classical" universe stops existing after the event horizon is passed.

I deliberately use the adjective "classical" because quantum laws still hold true inside the black hole, because it's space time independent. So, suppose that a quantum particle A is entangled with a twin partner B and B falls down in the black hole while A stays outside then A can still change B and B can change A. So, if A is observed by us and collapsed from the wave function then B is also collapsed. This is very strange, because in general it's impossible to read information from inside a black hole.

In fact observation via quantum entanglement in a black hole would be theoretically possible, but why can't we? That's because we need time to experience something and inside a black hole time has stopped. So even if we could interact somehow with the interior of a black hole by quantum entanglement we wouldn't notice because particle B left time. I mean, that the part of the universe where consciousness is still around, is left by B. So, in fact the interior of a black hole doesn't belong to our universe anymore. Information about particle B would have disappeared, which is against the law of conservation of information. It is supposed that particle B didn't fall in the black hole at all. It just got stuck on the event horizon as time stops there. Just as stars are, as they are being sucked in. All the information of all the objects that once fell in stays there. So, the event horizon is a history book of the black hole on a thin wafered surface. This is the same as a hologram. A hologram is a 2-D structure with information of a 3-D object.

If the hole history in 3-D is stored in a hologram can the universe itself be a hologram too? Black holes are non-approachable by space ships, so how would we know?
There's considerable circumstantial evidence that the universe is indeed a hologram.

Now, what about the role of (holo-)consciousness and black holes?
If you would pass the event horizon of a black hole with your space ship nothing happens. You would just slowly and easily enter the black hole and travel further.
An observer would see you "spaghettified", because your feet would be a million times heavier than your head over there in that extreme gravity field. Why is this discrepancy so extreme? The astronaut, being still conscious, doesn't notice that time is slowing down to the extreme. A second can last ten centuries and conscious experience is still placed in time for the astronaut. The observer sees the astronaut dying, torn apart as spaghetti. Because time comes nearly to a standstill the astronaut gets stuck in the last milliseconds of his life to stay there for eternity. This shows how time and consciousness are very relatee to each other.

Chapter nine

The theory of everything

Physicists are searching for the theory of everything. This should be a mathematical model in which reality is explained. At least, that's what they suggest.
In the previous chapters I have reasoned away all main principles of physics.
Time and matter don't exist. The universe is a hologram. Spiritual life has been put in the dungeons too. The soul doesn't exist. I have made clear that reality is an illusion too and that science misses (but needs) a more holistic view to make progress again.
Neuroscience is put to the hay stake, because our brain is just a localiser and not the generator of consciousness. Religion is sent to the underworld, because God, if existent, is only the conscious state of the universe itself and not a super being in contact with humans. I told you the philosophic views of consciousness, admitting that I belong to the realists, the gang of fools who assume consciousness is real and the only real thing existing.
We live in a scientific age and I am deeply in love with science. If science would be a religion, maybe the scientific world would persecute me now and torture me to death as an apostate.
Luckily that won't happen and only reasoning is used nowadays to bring me to other thoughts.
But my views are not so unique and I find myself accompanied by other free thinkers who have the same view. So, the reader is warned because the blasphemy to science will continue now, me being in love with science! What a paradox.

Now, "the theory of everything", the life work of Stephen Hawking and many other great modern scientists is actually a typical "physicist thing".
The assumption that matter is real and standing as a paradigm and holding out for so long is also causing the search for this theory. Physicists want to explain the interaction of particles within a universal system and hope to explain reality with that. Which they won't, because reality isn't made up by particles, but by immaterial phenomena. For physicists it's deep torment to realise over and over again they will not find out how to merge quantum physics and relativity theory, because they keep assuming that time and matter are existent phenomena and must rule out the role of consciousness, because they don't get a grasp on that.
But the great scientists who founded these important theories did give them a

hint! Einstein as well as Planck were attracted to eastern philosophy and even met gurus from the far East, pondering about the role of consciousness time, reality and the relation with the quantum collapse. Of course, this was a spare time exercise but that doesn't mean you can't take that serious.
So, I think, looking to particles alone for a theory of everything will be a dead end if you don't acknowledge it to a philosophical perspective.

As I proposed before, we have to see the universe as immaterial and consciousness as real. What is this (holo-)consciousness made of? The exact nature of consciousness is unknown of course, but one thing can't ever be ruled out and that's that the building blocks of conscious experience are at least made of information bits.
Not as zeroes and ones, but as a quality of universal experience whatever that may be!
To understand this view we have to get acquainted with information theory.

Information theory was founded by Claude Shannon, an engineer who got the assignment to research the noise on telephone lines. The mathematics he developed became the base for the behavior of information. He discovered that, just like energy, most laws applicable to energy are also applicable to information. Just like there's a law of conservation of energy, there's also a law of conservation of information.

All the information in the universe stays the same in quantity but is never the same in quality. Information is never lost, as I told you in the article about black holes.
In fact, we see the shadows of a "theory of everything" looming here, because if we consider the universe as the largest existing database then we can explain how this information is processed. First we need time. Time itself is, of course, also virtual information. Time has to sort out in which order events happen from cause to effect. It's entropy that states the direction of events. The natural laws determine what events are forbidden and what events are compulsory. (There's no in between). If there's time, there's also space. (theory of relativity)
Now what we also need is movement of particles, without movement of particles, nothing happens, so no events. A particle is an energy level obeying natural laws. The nature of energy is unknown. Both are the different versions of the same phenomenon. Energy is mass. ($E=mc2$)

The determination of events, in other words the quality or configuration of information in the database of the universe, is also directed by quantum laws making probabilities possible of certain states things in the universe or the whole universe is in.

As the universe is a database in a quantum state one might expect many universal states of one universe in many superpositions. I think this is what a multiverse really is. A collection of many, maybe trillions of possible universes that exist together at the same moment. The next moment one universe stays and the rest of the universes disappear. The universe that stays, is our universe. Until a new superposition of universes is reached leading to the same result, being again our universe. Now, this process is happening in the smallest amount of time, over and over again (Planck time).

What is making our universe ours? What's choosing what universe stays the next moment and which will disappear? In my perspective there's only one phenomenon in the natural world that's able to think and choose: (holo-)consciousness. So, it's the best candidate for that task.

So (holo-)consciousness collapses the quantum state of the multiverse in one experienceable universe, that's ours in a split second. So, our own (local) consciousness - being just a Platonic one, because of a brain processing it, isn't aware of this kind of universal processing of multiverses, we humans are just aware of the smooth flowing of events from cause to effect.

That's what brains are for. A good job.

Now, what main information does the database of the universe have?

These are categories of information: T is space-time, E is energy-mass, L is natural laws, the algorithm of the universe. The grand processor is holo-consciousness.

Now the philosophical implication is that we have to stay in this *"chosen"* universe, because (holo-)consciousness creates it as suitable for us and it's the only one with consciousness, the right natural laws for life and thus the only one that can be experienced. Al the other trillion ones are lost because they don't have consciousness in it or the wrong natural laws.

Chapter ten

There's no coincidence

Coincidence is an event that can't be predicted in advance,
You can guess, after all my philosophy in the previous chapters, that my view is that also coincidence is non-existent.
The definition of coincidence holds our insufficiency in it to predict an event, nothing more, nothing less. There's been a time that one believed that God caused all events. When religion was more and more replaced by science, a time came that everything could be predicted by physics. In Newtons time all-natural laws were mapped and it was obvious, when you knew the exact state of something, one could predict the next event. This holds for things like moving objects, celestial bodies, a pendulum and later on for molecules and chemical substances. This was called deterministic science.
Indeed the Newtonian science still holds in these cases, of course.

Take astronomy and space travel. We can predict exactly were planet Mars will be in the 3th of February 2045 at 12:05 GMT. We can send space crafts with a stunning precision to other planets sometimes more than one light hour away.
But can we predict were Mars will be in the year 502045 on the 3th of February at 12:05 pm? Certainly not. Because how small events in the solar system may be, they are there and they will have their effect in the long term. For example, our own moon moves away from us and the center of gravity of sun, earth, moon and Mars will change slightly and hence Mars' position in the solar system. And this is just one of the many things that will happen. So, in the long term even the deterministic science is inadequate of making robust predictions. So, the clockwork universe of Newton is not so perfect after all. In fact, everything is chaotic in the long term.
The era of quantum physics introduces total unpredictability again. In quantum physics anything can happen what isn't done by magic, but can be done by natural laws. So, complete unpredictability is back. With this you would argue that indeed everything is unpredictable, so coincidence is the main tune.
In a superficial view that seems to be so, but not when we go deeper into the subject.

If NOTHING would be predictable, cause and effect wouldn't exist either.

Cause and effect are really a "time and entropy thing". But also, a matter of consciousness. After all we experience things to fall down and not fall up; droplets of water splash on the ground and don't fall up to return to the droplet of water again from the splashing. If we rewind all events of the falling drop of water back in the film device that's no problem for the device at all, but for our brain this might be a very strange order of events. We can't cope with that. That's because our lives flow from birth to death and from young to old. Nature is built like that and we can't escape from this order of events.

Even, if we could stop time we wouldn't like being in a world where everything stands still and nothing happens. This would be a meaningless situation for us. The right flow of events is essential for us. Our price for that flow is coming ever closer to death, but we can live with that. So, time being an illusion, is special because of our consciousness, but a machine like a projector, being a non-conscious thing doesn't care about the order of events from beginning to end or the end to the beginning. But for us, the whole rewinding of the story of the falling droplet would be confusing, useless and meaningless. The order of events from cause to effect isn't just because of entropy alone, but also because of its meaning for our minds. So, we can only understand chronicity, which doesn't mean at all that the world really is synchronized at all. We just assume it is. Our subjective experience is convincing us of that.

In quantum mechanics time is in fact non-existent. Any state can exist at any moment. But this would mean everything happens at once and therefore consciousness is needed to read the events in the right order. After the collapse of the quantum wave function, time is consumed from cause to effect by conscious experience.

What about predestination and coincidence?

The fact that you read this is a very small cosmic chance indeed. First, you have to be born and being interested in reading this.
Your birth was dependent on our father and mother, grandpa and grandma etc. You must have bought my book. All in all, taking all these events into account the chance you reading this would be astronomical small.
But that's not completely true.

You can also reason that the chance to read this can be calculated by almost 100% if you have the proper information. All the prior events can be predicted if we just know exactly *everything* influencing these events.
How big was the chance of the meeting of your father and mother?
If your father and mother lived in the same village and loved dancing and had about the same age there's a good chance, they met each other on the dance floor and fell in love with each other. Now how large was the chance of "you being you" as one of the 40 million sperm cells of your father? Your genetic makeup might be unpredictable, but if you exactly know how genes do their job in the process of conception you could make a good scientific guess. All other events from your birth to the moment, you read this book, should be known too.

Here comes our next problem: we can't predict human behavior exactly, because it's highly irrational. So, we would need mind reading machines to predict that. Which minds do we have to read then? All minds that have influenced your life story from birth to reading this text somehow - directly and indirectly. These are probably many, many thousands of minds. But those minds have been influenced by many many thousands of minds again, which, etc etc.
Probably the whole population of the planet has influenced your life somehow very indirectly, if you want a very exact prediction of how your life will run from birth to reading this text. But even if mind-reading machines would exist, how can you predict free will of all these minds? What do they decide? Why and how?
Because even if you know word by word what somebody thinks, you can't exactly predict what someone's next action will be.
With a huge effort you could at probably predict for more than 50% that one day you will read this text. A very weak statistic chance indeed, but more than throwing up a coin. For us, sentient intelligent beings, knowledge just has its limits.
Unpredictability will always rule our lives, even with mind reading machines. Unless there's something omniscient and I suppose (holo-)
consciousness knows all thoughts of all existing minds because it is the mere source of it. I will come back to that later when I explain free will.

What about chaos theory and coincidence?

Chaos theory finds its origin in mathematics. Fractals are a product of chaos theory. Beautiful shapes that are manufactured by computers out of simple formulas. The most fascinating of these, is that nature's geometry consists almost entirely of fractals. The form of shells, tree leaves, spiderwebs etc etc.
Lorenz, the founder of chaos theory once spoke the famous words: "The wing beat of a butterfly in the Amazon can cause a hurricane in the Midwest (USA)".
Let me explain this again, but not with the example of an ant that is seemingly accidentally trampled to death, but this time with the position of the planet Mars in relation to earth over 500,000 years, as I mentioned in the beginning of the chapter..

In 2015, a space probe visited the asteroid Ceres, a large boulder of several hundred kilometers in diameter which also goes for a mini planet (asteroid).
You will say: so, what?
But with this action we have substantially changed the position of Mars in relation to the Earth over maybe 500,000 years. The spacecraft circled around Ceres for months, took pictures and was deliberately smashed on the surface. A waste of money, an act of indifference from NASA?
No! Due to the orbit around the Ceres space probe, Ceres is pulling at the space probe with its gravity, but the space probe also pulls at Ceres. Although the space probe is in disadvantage in terms of weight, the asteroid will still make a small

rocking movement of at most a few inches due to the attraction of the probe. Eventually, the orbits of adjacent asteroids will change slightly after a while (for example a few thousand years) and finally the whole asteroid belt will start to move a little differently (after e.g. 100,000 years). Now the asteroid belt lies between the planet Jupiter and Mars. The mass of the asteroid belt in its totality is substantial in relation to the masses of these two planets. For Jupiter, the heaviest planet in our solar system, this will not have much influence, but for the small planet Mars it will. After about 500,000 years, the orbit of Mars will have changed considerably by the movement the asteroid belt. The visit of the Dawn spacecraft to Ceres in 2015 was the cause of this.

It's out of the question that half a million years later is relevant to us, but it's obvious that trivial, seemingly unimportant, events (the wing beat of a butterfly) can have enormous consequences for nature (a hurricane in the Midwest USA) in the long run.

The decision to crush Dawn on Ceres is based on the chaos theory that predicts that the orbits in the solar system will be disrupted if Dawn were to continue to circle Ceres forever. You should also consider self-interest. The asteroid belt occasionally loses an asteroid that sometimes passes very close to the earth, the so-called earth near earth shearers. A disturbance of the orbits in the chaotically moving asteroid belt may well result in new earth shearers and every earth shearer is one too many for our civilization (see the film Deep Impact). This decision by NASA is not indifference but is calculated rationally on the basis of chaos theory.

In my book I use a few other examples that have not so much to do with mechanics, but with human choices. Hitler might have won the war if he had not stopped his tanks at the Battle of Dunkirk to make the British retreat possible, but his full bladder prevented him from thinking clearly. The fact that there are no swastikas on the government buildings in London is thanks to Hitler's bladder.

How about the universe and coincidence?

You could say that the most of the mechanics of the universe runs on the "autopilot", i.e. the laws of nature are sufficient to make the universe work. The laws of nature are mainly described by mathematics and are highly deterministic. By this I mean that every cause has a consequence according to the prediction of that law of nature.

It is assumed that the blueprint of the universe was already ready there at the very moment of the big bang itself and the discovery of the background-radiation proved that this was correct. The background-radiation is the imprint of the big bang with zones of more or less energy (information density) which are also assumed to be quantum fluctuations. In the places where the amount of matter was dense enough stars and galaxies arose.

"Order" in the universe is not quite "orderly" as I explained in the previous text with the orbit of Mars, there is always an element of chaos in order (chaos theory), but also these chaotic processes also behave according to the order of natural laws.

The second element in our universe is of course "chaos". Chaos makes any prediction impossible and in a chaotic system every meaningful and coherent cosmos is excluded. Our universe is in its foundation a quantum universe and therefore naturally completely chaotic, bizarre and elusive. Yet we experience a coherent integrated universe, how is this possible?

As I explained in "the theory of all " chapter, the number of universes is in principle innumerable (but not infinite) and of them, only one universe remains where conscious observation is possible. The quantum state within that unworkable chaos, with its elusiveness and unpredictability can only be eliminated by the quantum collapse. It's holo-consciousness which is doing that. Holo-consciousness thus creates, from endless universes, one universe in which everything is just right. The only universe of all those innumerable quantum universes (the quantum multiverse) is where everything is exactly right and is our own familiar anthropic universe.
(Andros = human, Greek)
A universe which has consciousness and where observation is possible is also called a Goldilocks universe , which is an allegory of the fairy tale of Goldilocks and the three bears finding her sleeping in a small bed, that exactly fits her body size.
If this "chosen" universe is the outcome of the quantum collapsing of numerous other quantum universes it's obvious that all events in our universe must be exactly right to build this unique universe. Nothing, really nothing, can be left to any coincidence from big bang to heat death of the universe. It has to be an absolute perfect sequence of perfect events.

What about consciousness and coincidence?

Having said this, we come to the third phenomenon in our universe that excludes coincidence and that is "consciousness."

Consciousness is the largest variable factor in the universe. After all, consciousness is capable of making choices and that is not the case with natural laws (here everything is fixed, it is like a clockwork) and also not with quantum mechanics (here nothing is fixed, there is only coincidence). Making choices depends on a large number of complex factors that work together or counteract

each other. Our whole evolution is an evolution of successive choices of sexual partners, of preferential chemical reactions in cells, of specific genetic feedback, of biological balance and so on, and so on.
Every living creature thus makes its choices in its own way. This applies to viruses, single-celled organisms and humans.

Even a bacterium that migrates to a glucose-rich environment makes a kind of choice, however simple and automatic. That this choice is not driven by intelligent conscious action is clear, but strictly speaking a falling stone can't choose whether it falls to the left or right and a bacterium does. The fact that the behavior of "primitive" life is not as complex as the behavior of a person doesn't mean that there is no behavior at all with bacteria. All systems that make autonomous choices have a certain degree of programming and information.
I explained earlier that all information that exists is present in the holo-consciousness.
The movement of a single bacteria is not merely accidental, but staged on the basis of biological programs. Also, in "primitive" cells there are programs and routines that do not happen by chance but according to certain laws that make sense, having a function and a goal.
So, it's not always complex conscious decision-making that rule events, but mostly simple almost invisible processes.

What about the coincidence we, conscious humans, experience?

Are our choices based on mere coincidence? Our choices are far from accidental and are linked to information processing in the brain. The number of choices that a single human being makes in his life is already countless, let alone the sum of all choices of billions of people. We have, despite this enormous variation, more or less control over our fate, apart from natural disasters.

We always have the choice to destroy ourselves by developing a climate disaster or a devastating nuclear world war or surviving by adaptation and rational choices.

If we send a satellite to Ceres, the orbit of Mars will change. If Hitler had not had a bladder problem, he would have won the Second World War, because then he would have conquered Great Britain and the invasion of Normandy would have become impossible. The bladder problem of Hitler is no accident, but ignorance, because the bladder problem was dormant long before the battle of Dunkirk. We simply miss this information. Or did Hitler find it too embarrassing to discuss this with his doctor? So, you must BE Hitler to reason this. The only one who can know what "it's like to be Hitler" is Hitler himself. So again, here's our information gap and our dependency on consciousness.

Einstein said these wise words after the discovery of quantum physics: "Der Lieber Gott doppelt nicht!" (Our Lord doesn't play dice!")

Our choices in our lives may already be fixed from birth to death. The fate of humanity might be too. Eventually everything is basically quantum and changed into the only possible universe already completely written out in space and time, like a book or a film. It is the best and only option and any other scenario is inferior. We live in a chosen universe.

The universe can't play dice. It must be absolutely perfect. Everything must be experienced in an already determined scenario which was instantly made out of numerous possible quantum universes (quantum multiverse).

Time is an illusion, it's created by consciousness, present past and future exist at the same time.

Coincidence doesn't really exist either and is just our ignorance. Consciousness reads us the story in the right order from cause to effect and makes it meaningful. Holo-consciousness is the grand cosmic processor. It can't make coincidence spoil its story.

Quote Albert Einstein:

<If this universe were the result of a blind coincidence in its million-fold order and precision, then it is just as credible as when a printing office explodes and all the block letters come to the ground in the completed and error-free form of the dictionary.>

Chapter eleven

The absence of free will

Although the experiments of Libet, a neurosurgeon, who, in the sixties, experimented a lot with patients during a brain operation by giving them instructions, are controversial, no one can deny that they have shown that we shouldn't think that we control our own brains completely. Brain research has many limitations.

First, our understanding of consciousness, free will and ego falls short, and secondly, the inaccessibility of our brains for research.

An advantage of brain surgery is that the patient doesn't need an anaesthetic and can stay fully conscious. Brains are insensitive to pain. With electrodes you can stimulate the brain and then ask the patient what he experiences, the surgeon knows which location he is approaching. Libet discovered that if the test subject independently carried out an assignment, the brain performed this earlier than the patient experienced it. Then Libet stimulated the location where the assignment came from, but the patient experienced it as if he were executing that assignment voluntarily. Libet further refined his experiment and gave the test subject control over his own brain stimulation via a button. He was told that if he pressed the button his arm would move. When the test subject did this, he strongly denied having pressed a button, but claimed in full conviction that he moved his arm spontaneously and without the intervention of a button. In addition, Libet discovered something shocking: a conscious experience only takes place a half to 2 seconds after the current event.

In other words, if you see a bird flying, it actually flew there 0.5 to 2 seconds ago! Our consciousness is constantly lagging behind reality. Libet's conclusion went a long way because he claimed that his experiments showed that the brain takes all the decisions and then lets us experience the illusion that we do everything of our own free will. Moreover, we are only a little bit behind reality, let alone that we would still be able to say something about it afterwards. There was a lot of criticism for Libet. Yet, his work has been invaluable to neuroscience. Especially from the fields of religion, psychology and philosophy came a lot of resistance. Yet, his experiments were confirmed time after time.

Your author believes that free will is primarily an illusion. That doesn't mean that it is not an unimportant illusion that we can't work with. As long as we, ourselves assign free will, we consider each other responsible for our actions and it's me who does or says something and not my brain, and I'm responsible for my actions.

It doesn't matter that it was actually my brain. That is an academic discussion. Criminals should not go free with the argument that their brains robbed the bank and not they themselves. For the religious man, the devoted creed of free will is a condition to be rewarded in the afterlife. I certainly wouldn't want to deprive that religious person. Although your author largely endorses Libet's conclusion, he doesn't believe that we actually function as biological robots without our own will. Our mind is connected to our brain body system just as atoms are connected to energy, mass with time. It can't exist without each other. No functioning person without consciousness. No consciousness without body. No life, no consciousness. Robots will never be conscious. Without the illusion of free will, no meaning of life. Without atoms no functioning universe, without mass no time. And vice versa.

The famous "Zombie discussion" illustrates nicely how delicate an illusory free will and consciousness is. Suppose that half of humanity consists of Zombies, would we then be able to separate the Zombies from the conscious people? There are no consciousness measurement devices and we can't differentiate our brain activity from theirs. Zombies could move unnoticed among us. We could fall in love with it and Zombies could be elected as president without having the slightest notion. That way Zombies could reproduce. But would Zombies know which partner has consciousness and which does not? And how would the genes behave? If the Zombie gene is stronger than the consciousness gene, then the fate of the conscious man is lost and the fate of the Zombie is confirmed. If they are equivalent, they remain in balance. And would that matter? This question does matter. Namely, the fact that we think that a conscious being has different rights than lower life forms, let alone objects without consciousness. We give our pets a lot of rights, but a laboratory animal, insect or plant almost nothing. Yet they are all living beings. A stone is a thing for us. A Zombie is admittedly a living being but without consciousness. It is no more than a stone. Zombies therefore have no rights and are just a thing.

So, our relationship between our fellow men whom we love or choose as leader is determined by this. After all, you do not have sex with a stone and you don t choose it as president.

The same problem occurs with artificial intelligence and in a not too distant future we will have to deal with this. People-like machines will start copying and using our behavior so well that they will no longer be distinguishable from the real thing, just like the Zombies. Does the robot get rights or does it remain a thing with fewer rights than a living creature?

It doesn't seem likely that robots will ever have consciousness.

Sir Roger Penrose wrote in his famous book "The emperor's new mind" that he could prove that this is impossible. Another mathematician, the genius Alan Turing, who deciphered the enigma code in WW2, was convinced of this. He developed a test that could distinguish between conscious beings and artificial intelligence.

This famous Turing test, in simple terms, means that the machine and the human being can't see each other and ask each other questions about what they are. A human has always succeeded in knowing that distinction with a few questions within a few minutes, but a machine could never do that. Even pre- schoolers are successful in the Turing test and even an adapted version for chimpanzees was successful for the chimp. If the Zombie has human brains, but does not have any conscious experience and does not know that he exists, will he pass the Turing test?

I certainly don't think so. In our thought experiment with the Zombies, we could effortlessly end the interesting discussion by eradicating all Zombies without breaking the law or have grief or conscience about it. They do live, but have no spirit or soul. You can destroy that with impunity after you sorted them out with a Turing test for detecting Zombies.

Goodbye to your beloved or that great popular president!

Humanity can't afford to lose consciousness through genetic selection. Knowing that there are no consciousness measurements-devices and no Zombies or consciousness genes, we must leave the Zombie-discussion for what it is and focus on the problem of free will, but the thought experiment shows how much value consciousness has for us.

I have already indicated that I think that free will is largely an illusion. I don't find Libet's experiments sufficiently conclusive to totally exclude free will.

I distinguish three types of will:

- the reflective free will

This is the will that is needed to survive. In an emergency (e.g. an accident) where a decision must be taken within a fraction of a second, the inertia of the consciousness is not adequate and the brain is on autopilot. This is especially what Lebit has studied.

- the operational will.

This is the will of everyday life. "I want to carry out that specific task today, for example going to work, completing the report, attending that meeting". The awareness is operational here, but many tasks are carried out unconsciously, such as driving to work, typing on your laptop, hanging your coat on the coat-hanger, etc. This is a mixture of the reflective will and the free will.

- the existential will.

There are no unconscious elements here. The existential will is permanently present in consciousness. It is your "drive" and is linked to your personality." I want to be the best, I want to be the prettiest, I can do it better, I have to prove myself "and so on.
You can't escape this kind of intentions and striving, you always commit yourself and you are constantly aware of it. The problem with Libet's experiments is of course that the tests in his operating room couldn't tell anything about the operational will and existential will that have a real meaning than just lifting an arm.

It's clear that our brains limit our free will to what doesn't concern our survival or daily operations. Lebit also showed that everything our brains do is labelled as something we do out of "free will", while that is usually not the case. To a large extent what we experience as "free will" is mostly a persistent illusion which is generated by our brain.

Although the reflective will is well understood from the theory of evolution that strives for the survival of the species and the operational will is to be understood from the behavioral psychology and neurosciences, the existential will remains a mysterious phenomenon and perhaps even holds a spiritual character in it and in our language, we call it our" soul".

Chapter twelve

Quantum consciousness

I must admit that I admire you, that you made it to chapter twelve, because all the subjects I spoke about are claimed not to exist, so what am I talking about anyway?
Matter doesn't exist, information does exist as well as consciousness. The soul and free will don't exist and the universe is the outcome of a quantum multiverse where anything already happened in a split second. Moreover, nothing is coincidental.

Consciousness is a concept we can hardly deal with and quantum science is so strange that almost no one can understand it. Writing an article about quantum consciousness is a tricky business, but I'm going to try it anyway.

In the first instance, we must realize that reality is not a well- defined concept.

Reality depends on three global levels of understanding:

1. What we perceive is the Platonic reality. Our everyday experience of reality.

2. That which can be proven, even without observation, that exists is the natural reality. This is the reality that can be demonstrated not only by observation, but also by measurement. But you need proof for it and thus science. For example, the colour yellow is observed of course, but can also be measured as an electromagnetic vibration with a specific frequency.
A black hole can't be observed, but the evidence that they exist is too overwhelming that it can't be denied that it belongs to reality. Atoms can't be perceived, but we are certain that they exist.

3. That, which we can neither perceive, nor measure *directly*, but only reason to exist by means of indirect evidence (e.g. with mathematics and physics) is the quantum reality or the super reality.

As the name suggests, the order of these three layers of reality is not chosen by chance, because they say something about the depth of reality experience and its abstraction. After all, Platonic reality is the superficial reality of everyday life that we take for granted, natural reality is well studied by science and learned at school (already more abstract), while super-reality is something that is only studied by specialists, such as physicists, cosmologists and mathematicians. (very abstract)

Why is this important?

Consciousness is not something that is only between our ears or is only present in our skull. Consciousness is a universal information processing system that is not exclusively made in the brain. It is everywhere. At least that is the conception of the philosophical mainstream of "realists" who are studying consciousness and I defend that view.

I mean holo-consciousness with that, because this is all consciousness and information that exists and is therefore everywhere. How that universal consciousness can be operational and maintain a working universe remains a mystery, but our own consciousness is too. We know our consciousness as something that is there day in day out and our goal is to maintain it as long as possible. After all, it is our whole being. We call this "our life time".

Holo-consciousness is therefore something we can't perceive, nor understand and even can't imagine a phenomenon which holds all information of the cosmos and doesn't have a physical body, nor a brain, which we hold as an absolute condition for our own consciousness. Holo-consciousness is so abstract that we couldn't even draw it or find a mathematical formula of it. It must be at least in the third layer of reality but actually is more than that, it's all reality!

Another solution is to maintain a theistic concept and to say that holo-consciousness is the same as God. The philosophical problems that arise then are that holo-consciousness returns to personal form which has been proved to be wrong.

It is important that we realize that we too are unquestionably part of holo-consciousness even though our ego tells us the opposite. After all, I am in the "here and now" and I am not integrated in the rest of the universe.

Holo-consciousness is in all layers of reality, the Platonic, natural and quantum reality. After all, it is all there is.
Holo-consciousness is in me and in the rest of the universe.

However, the study of the brain tells a completely different story. It is regarded as an imposing edifice of neurons with trillions of connections that senses perception and is able to make a picture of the Platonic reality.

It was again Sir Roger Penrose and David Hameroff who pointed to the possibility that our brains might also work as a quantum computer. Their theory states that we also perceive through the quantum system.

When we look at a flower for example, the light of the flower falls on our retina and through a complex brain network an image of it is formed in our optical cerebral cortex. This is how it works according to the principle of natural reality. But nobody understands exactly how that image is formed and how that becomes conscious.

In the quantum world it works completely differently, but that goes completely against our intuition.
The flower is not perceived by our eyes but is the quantum state of a flower but due to our perception (measurement) of the quantum state it is collapsed to a well localized flower and then can be observed by sending photons with information of the flower to our retina and then translated to "brain language" or neurons firing a pattern or starting up different neuronal networks.
The information about the flower exists somewhere in the "all- encompassing database" of quantum reality, but is only converted into material reality when a perception of the flower takes place by a conscious being. There is not a real image of a flower in our optical system at all, but there has been a quantum collapse that has led to the image of a flower.

Brains only translate the result of that quantum image into the context of a flower as your personal experience, because it recognizes certain neuronal networks confirming the flower concept.

An important argument for the brain as a quantum computer is the slowness of the neuronal system. In our nervous system, stimuli have a speed of a few meters per second, which is such a snail's pace, that smooth complex processing can't be fast enough to make our daily interaction with the world adequate.
The quantum world is instantaneous, i.e. images and impressions are there immediately without loss of any time, without having to undergo the difficult neurochemical delay via a nervous system with synapses. Neuroscientists reject this view and point out that the development of the optical system was an evolutionary advantage and so must be the only explanation of the experience of seeing a flower.

The defenders of quantum consciousness don't see that system as superfluous, but as a necessary check-up of the image that the quantum collapse has formed and is therefore more like a confirmation afterwards decoded in "brain language". So more clearly, the localization and imagination of the flower is instantaneous and generated in the third layer of reality, the quantum world. The brain gets to link to this image by generating the right neuronal networks which recognize it as a flower. Of course, the conscious experience is only complete when both processes are linked. That confirmation is experienced as the Platonic reality. That explains why our consciousness lags behind reality for a few seconds.

Of course you could oppose this view by pointing to the other senses which are not depending on observation of photons like sound, touch, smell and taste. The defenders of quantum consciousness defend their view by stating that sound (the movement of air molecules against the eardrums), pressure (the deformation of a sensory apparatus in the skin), the chemical reaction of sensory units in the nose and tongue also have quantum processes in it.

Sir Roger Penrose and David Hameroff couldn't prove the quantum consciousness theory, although they have attempted to make the so-called microtubules (wires in cells that are a single molecule wide) are responsible for the transmission of quantum information. Something that is very controversial and I also find very artificial.

Summarizing, we can say that observing a flower first conveys the information about the flower from the quantum world via the quantum collapse to the natural reality (a collection of molecules in spacetime) and our brains, which can then perceive the material flower through our eyes., then bring the flower as a meaningful phenomenon to our Platonic reality. A conscious perception thus traverses all layers of reality and finally synchronizes the three realities into one total reality.

The information holder of that one reality is the holo- consciousness, I am an integral part of it, just like you are, the reader.

This also explains something about being alive and dead. In Platonic reality, dying is the transition from a living being to a corpse. Human remains are only a collection of molecules, a material thing, while a living human being, who also just exists of molecules, is in our eyes a person.

You could also say that dying is the transition from Platonic reality (which requires brain activity) to natural reality (the world of things, matter). The three stages of routing through all sub realities is now gone and therefore also personal consciousness.

Finally, the mortal remains will also decay, but the information about your life (all conscious experiences of your brain body system) can't possibly be lost, even if your brain has already been dissolved into some carbon atoms. After all, the universal law of conservation of information prohibits that. Where is all that information about your life then? All information is retained in the holo-consciousness, where it is timeless and space-less, back to the third layer of reality!

The resemblance with the religious conceptions about the immortal soul, which leaves the body and enters into an afterlife, now imposes itself on us. Although I do not share this view, I can imagine this as a concept.

The discussion whether quantum consciousness really exists is far from being completed and has perhaps not even begun, but I find this explanation for the nature of consciousness far more elegant than to claim that consciousness is formed only by small electrical currents in brain cells, which are much too slow to even make thinking adequate.

Though the quantum information processing is very efficient and quantum computing will once support our daily life, it's unlikely that within the brain, quantum computing takes place as Penrose suggests. For that it's too large, too wet and too much disturbed by other noise. Penrose microtubules aren't needed for quantum computing as the universe primarily is already in a quantum state. The imagination of a flower is everywhere in the universe and therefore also nowhere visible in that state. That's the consequence of the quantum weirdness. The brain just "pulls on the right drawer" of the universal consciousness to experience a flower.

Chapter thirteen

Evolution despite Darwin

There is no reasonably thinking person in the 21st century who seriously thinks that God created heaven and earth in seven days, perhaps except for a few religious fanatics, though a large population of orthodox Christians in the USA still hold this for true.

It is the great merit of rationalism and the scientific method that humanity started to realize that we're not in the center of the universe, but rather a rare phenomenon on the edge of an inconceivably large empty space.

Darwin lived in the Victorian 19th century when the British empire was spread over all continents, but despite its greatness, the petty spirit of its zeitgeist was also very common. It was therefore very brave of Darwin that he published his "Origins of species", when he returned with the Beagle from his journey around the world. He was immediately the center of ridicule and scorn and for some time the outcast of the scientific world. Let alone the religious world! Nowadays, the theory of evolution is widely accepted and the evidence that evolution has taken place is so overwhelming, that this theory has become a science and a hard fact.

Well, don't presume that I am an admirer of Darwin. I admire his courage and accept that his publication made him the founder of the theory of evolution, but I think he gets too much credit. He was certainly not the only originator of the evolution theory.
There was also Russell Wallace, who did a lot of research especially in Indonesia (then the Dutch East Indies) and devised the theory of evolution there. Only Darwin, with whom he had close contact via correspondence, published his work earlier. Darwin therefore does not belong to the list of super geniuses such as the physicists Newton, Einstein and Bohr.

Yet, some scientists would rather see "The Origin of Species" on their bedside table than "The Holy Bible". I always find this exaltation of Darwin exaggerated. Special Darwin days and all sorts of festivities designed to honour the great bald man with the long beard, but why actually? Because he gave a hard blow in the face of the biblical worldview in the context of atheistic ideas? He wasn't he the only one, was he?

Darwin and Freud belong indeed to the same category.

Freud was also a great man just like Darwin. Yet, they were both wrong about how their theory worked, but they certainly were both founders of a new science. Darwin of evolutionary biology and Freud of psychiatry.

Darwin claimed that natural selection was the motor of evolution and Freud that human drives in the subconscious were suppressed by a conscious superego. Meanwhile, we discovered Mendelian inheritance (Neo-Darwinism - how could it be otherwise?), DNA and RNA molecules and finally epigenetics.

Neither Freud nor Darwin are to blame for those later scientific discoveries that brought other and new insights.

What is to be deplored is that the evangelists of Darwin continue to proclaim his gospel against better judgment. The idea that evolution is driven by purely sexual or natural selection has long been refuted.

Yet, you can't see a film about nature on the media or you hear a thoroughly Darwinian commentary.

An impala that has strayed from her herd is being chased by a group of lions and eventually killed by them. The comment is of course that the weaker specimens do not survive. Because the impala happened to stumble, she was weak or was she just very unlucky? Or in a TV quiz the right answer is "that women find men with tight buttocks attractive because they would provide better offspring"! Where are the scientific underpinnings of this kind of nonsense? How dare you include this in a quiz!

That's marketing, not science. It shows again Darwinism in all its ugliness.

In a forum about science and philosophy I was participating in a few years ago, emotions ran high as soon as one had any doubt about the correctness of Darwin's theses. Suddenly, you belong to the suspicious camp of creationists (those who still believe that everything was created by God) or to the dangerous sect of the "intelligent design" movement. A clear and critical scientific view of evolution is apparently not possible even in the 21st century, not even in the well-educated community that should know better.

Good old Darwin ... what if he knew his theory of evolution was in a mechanistic way turned upside down? I'm sure he would agree.

1. The discovery of DNA by Watson & Crick. (Both Nobel Prize winners). Not the natural selection by animals, but the expression (*) of genes stored on a phenomenally large molecule in the nucleus of animals and plants is responsible for the dynamics in evolution.

2. The discovery of DNA methylation and regulatory genes. Genes can be turned on and off. With this, life adapts.

3. The discovery of epigenetics. The influence of the environment changes the expression of the genes and this expression is inherited for many generations.

4. The discovery of the microbiome. Viruses and bacteria from the environment change the genes of animals and plants.

5. The discovery of the evolutionary leaps. Sudden changes and accelerations of evolution are the rule rather than the exception in evolution. It is more a repetition of revolutions than a slow evolution, as Darwin thought.

6. The discovery of extinctions. An extinction (*) leads to a new even better version of evolution.

All in all, there is little left of Darwin's original theory, that sexual selection and natural selection or spontaneous mutations alone, are the motor of evolution.

Modern counter-arguments are:

-simple life such as single-celled organisms procreate asexual and can't operate from natural / sexual selection, but are the foundation of evolution and even life, they are just everywhere!

After 800 million years on our planet, there would only be single- celled organisms if there wasn't something special going on, why else would a spontaneous process organize itself into cooperating cells that could form a plant or an animal?

-spontaneous mutations will not lead to improvements in 99.999% of cases, but rather to degeneration or no change at all. Even if this would lead to any genetic evolutionary improvement, the effect would be "diluted" because the gene in the next generations is challenged by other inferior genes that dominate the genome.

This makes pure Mendelian breeding (*) impossible, which every dog breeder already knows for a long time.

- the lifespan of certain life forms has no influence whatsoever on the speed of evolution. You would expect that life forms that live briefly evolve much faster than long-lived life forms. In other words: the one-day fly has 36,500 generations in a century and is more likely to have spontaneously favourable mutations than a giant turtle that will turn 100 years old. But there's no difference in evolutionary speed.

So, adaptation is not mutation dependent, but environment dependent.
All these arguments and newly discovered scientific facts disprove the basic idea of evolution that it is created through time by *random* mutations and natural selection. The main objection to natural selection is that it is not the mutual struggle of species or individuals that determines the direction of evolution, but rather the mutual cooperation. Could oxygen consuming animals exist without oxygen-producing plants? Ecosystems are much more important than the species and species again more important than the individuals, the organs more important than the tissues, the tissues more important than the cells and the cells more important than the molecules. And so, the whole is more important than the sum of the parts, holism clearly is the principle of living nature.

How does evolution work? Let's start at the beginning.

First of all, we live on a planet that is liveable by three indispensable conditions:

  - the right distance to her star (the sun), not too hot, not too cold

  - the presence of fluid water not too much, not too few.

  - the presence of a protective atmosphere.

Life simply isn't possible without these specific circumstances.
Yet it took two billion years for planet earth to meet these three conditions.
From that moment on life was almost immediately present and almost a billion years ineradicable thanks to a favorable interplay of cosmic, geological and biological factors.

I mention a few cosmic factors: a big moon, which keeps the planet in a stable orbit, a large planet at a distance (Jupiter) that traps asteroids in the solar system, a medium sized star (the sun) that burns for a long time (potentially 10 billion years) and thus can provide energy for long enough evolution to take place and also has a position in the galaxy (the Milky Way) that is far away from cosmic violence of supernovas, black holes and quasars.

Geological factors are plate tectonics, the movement in the earth's crust, the atmosphere, the oceans. Biological factors are the ecosystems that are perfectly aligned. For example, plants produce oxygen and animals carbon dioxide that they breathe from each other.

Bottomline is that evolution is not possible without this interplay. It is cooperation between the planet, the cosmos and nature that makes this wonder possible.

We have, so to speak, an organic planet (Gaia theory). The planet harbors life and life change the planet and so on.

Let me give you a concrete example. Imagine a savannah where grazers live that are kept constant in numbers by the presence of predators. Suppose the number of predators decreases (for whatever reason) then the number of grazers will be so large that the savannah will be eaten and turned into a desert. The grazers migrate to another area followed by their natural opponents, the predators. So, the environment has changed through life. The desert will host new life such as cactuses, gerbils, camels and rattlesnakes. So, the planet again changes life here.

But what drives evolution to create camels and rattlesnakes? Not spontaneous mutations and natural selection, but epigenetics.
Certain genes are blocked by the inhospitable desert life in existing animals (for example the gene that leads to more excretion of water) and others are turned on (e.g. the gene that leads to water retention). The genetic pressure is then created by the rugged environment and is inherited by the next generations for as long as the desert exists.
So, it is not the biology that directs evolution, but the interplay between planet and life. Adaptation, interaction and gene regulation.

The single-cell organisms (bacteria and viruses) that have populated our planet for a billion years play an extremely interesting role. In this way that they also populate our digestive system and influence our body in such a way that they have a comfortable and stable host. The properties and genes of all these local microbes have been adapted for a long time to their environment and are therefore also favourable for the host. It suggests that they contribute to the evolution because the microbes from the desert adapted in another way then in the savannah. They also push their host to another genetic system because they occupy most of the total genome.

Finally, a word about extinctions. Large extinctions occurred where 99.9% of all life was wiped out. In all cases, the most basic information in the genes is mostly retained and even if only two fish remain in the ocean, they will reproduce and resume evolution rapidly, but with a fresh start and in a changed, not very competitive living planet. Without extinctions no revolutions and no evolution. Here too, there are not too many extinctions and not too few. Both would stop evolution.

In summary, evolution is the interplay between mother nature and mother earth. The native Indians surely knew it all along the way…But much research has to be done, despite the Darwin cult.

(**) Mendelian breeding is improvement of the species by the principle of Mendel's genetics. That is to say that dominant genes dominate the non-dominant genes (recessive).

(*) Extinction is the extinction of almost all life due to a natural disaster

Chapter fourteen

Spinoza's pantheism: God is nature

Baruch Spinoza, (1632-1677) belongs in my opinion in the list of Goethe, Kant, Nietzsche, Schopenhauer and Descartes.
Illustrious German and French philosophers who changed our world view.
Whoever has read my first book will know that I count him among the greatest.
His modest and thoughtful nature meant that his life's work the "Ethics" was discovered long after his death as a genius set of thoughts.

Spinoza was a mathematician, lens sharpener and philosopher. You can place him historically as a philosopher from the early enlightenment. A time in history where people openly doubted the correctness of religion as absolute truth. That is why Spinoza is placed among the atheists, while this placing is certainly unjustified.
In fact, you can place him with the theologians who put philosophy or free thinking above the Bible.

Spinoza considered God equal with nature. He did not see God as a person with supernatural power, knowledge and immortality, but he saw the universe as the manifestation of God himself.

He literally saw the Divine thoughts and imagination as a result of reality.

As a Jewish Portuguese believer, he quickly became a persona non grata and the zeitgeist of the time demanded that was banished by the synagogue. It is unfortunate that Spinoza, out of fear of further persecution, as the freethinkers of that time often experienced, published his book the Ethica after his death. This has hampered the dialogue with this great thinker and makes his intentions susceptible to speculation. The atheists of our time claim him as one of them, which is certainly not true.

Why is Spinoza so up-to-date?

In the last century, the basis for quantum physics was laid. Observation of elementary particles by conscious beings turned out to be the cause of the so-called quantum collapse, the phenomenon that a virtual particle or a quantum his is the smallest possible piece of information from the universe) can only exist if it is observed.

This makes consciousness crucial to the existence of the universe, after all everything consists of elementary particles and thus everything must be observed in order to exist.
So, there should be a universal observer who can observe the entire universe.

In my book I call it holo-consciousness, a metaphysical term, which means that it is a consciousness that is everywhere, perceives everything and thereby makes the universe exist.

The fallacy that our brains are the producer of consciousness seems to be supported in the perspective of quantum physics. After all, we can't perceive everything from our tiny planet, while the universe does exist outside our scope of perception. Our brains are, one could say, "inhabited" by the holo-consciousness, but doesn't generate consciousness itself. The brain is only the living hardware and not the software. This theory is supported by the philosophical movement that deals with questions of consciousness as the mainstream of "realists", which I explained earlier. This is the view that reality doesn't consist of matter, but is an illusion created in the holo- consciousness. So, again: "Mind over matter".

Spinoza had these insights already in the 17th century, while he obviously couldn't know about quantum physics.
It's no surprise that other great minds as the physicists Einstein, Oppenheimer, Feynman and, to a lesser degree, Bohr endorsed the philosophical views of Spinoza.

The oldest known religion is the natural religion that holds the conviction that everything is "animated". Pantheism (pan = everything, Theos = God) is actually the most fundamental philosophy of all religion.

An example of such a religion of nature originates with the Indians, but even with prehistoric humans. They believe that everything is animated. Although we now know that a central nervous system is necessary for experiencing consciousness or a soul, this is true in a wider perspective, after all nothing is material and consciousness is everywhere. Pan psychism, the doctrine that everything material has a form of consciousness, is not very popular among philosophers and scientists, but in a broader perspective there is a core of truth in it. After all, if matter is only information, then every piece of information would be part of a larger amount of information, and so forth, which brings us back to holo-consciousness.

It's a fact is that pantheism actually forms the basis of all religions that followed later, Hinduism, Christianity, Islam and Buddhism. In all these religions the consciousness of a supreme being is the central theme.

Spinoza was horrified by the idea that God would personally deal with the trivial ups and downs of you and me and also claimed that the prophets were definitely human messengers and not, for example, the son of God, as the Christians claim.

Islam sees Mohammed as a human being and that also applies to Buddha. Eastern religions also see reality as "Maya" an illusion that exists only when there is universal consciousness, which is not only between our ears, but everywhere.

Spinoza denied that God would have something to do with good and evil. He thought ethics was a human invention that was "nootwendig." (Old Dutch for necessary).

In my first book I distinguish chaos and order as the two major components of the universe and their interaction causes complexity. Nature is complex. The living nature ultimately leads to consciousness through evolution as we know it in animal and human form.

Holo-consciousness is of a completely different order.

After all, it encompasses all the information that exists in the universe, and even time and space are just information and thus are totally abstract.

Spinoza therefore, saw God as an abstract concept that could not be imagined. This corresponds with the views of Islam and Eastern religions.

That the evolution on our planet could eventually lead to intelligent consciousness is a logical consequence of complexity (the interaction between chaos and order), but does not have to be a goal in itself. Although we can't determine that as a fact.

The better and longer complexity can develop, the greater the chance of the emergence of intelligent consciousness like ours.

In the vision that the universe is going through its own learning process, the vision of an "intelligent outcome" model as a product of evolution and the 'proper working universe" is a precondition for the only universe that can ever exist and be experienced. A quantum multiverse that selects the most intelligent outcome from trillions of possibilities. Choosing is, after all, only a property of consciousness.

Complexity always demands choices that work best in the organization of nature.

Most scientists and also most atheists are tenderly nicknaming nature as "mother nature" or mentioning the "self-organizing ability of nature or universe", because they can't deal with a world that self-organizes.
Spinoza would, had he lived in our time, have been a celebrity and a world-famous philosopher, a superstar, who would visit all talk shows on TV.
He died in 1677 as a simple lens sharpener in the Dutch town of Rijnsburg, but he is undoubtedly the most visionary theologian / philosopher we have ever known.

What developed earlier, the bird or the plane? The eye or the camera? The paw of a gecko or the klittaband? The chameleon or the military camouflage suit? The plant leaf or the solar cell? And so on, and so on ...

We can't escape the thought that our great intelligence pales to that of the intelligence of nature. The list of brilliant constructions of nature is immense and we have only studied and understood a fraction of it. It is in particular this built-in intelligence that makes many people skeptical about the theory of evolution. Yet there is no doubt that evolution has taken place. Creationists insist that God created nature. The intelligent design movement says that God came up with the designs and guided the evolution from phase to phase. Regardless of your view, as a scientist or religious person, it is naïve to think that the product of evolution is only a succession of stupid random mutations that just by chance turn out to be very ingenious.

Although no sensible thinking person should think this, it is what many scientists still assume and I have great difficulty with that. It seems like a frenetic denial of something that would lead to something much greater than our own mind, in the worst case to the mind of a God!

Back to the previous topic, the evolution.

A giraffe was once a horse that developed long legs and a long neck because the foliage in the area hung too high to reach. The information on how your legs grow extra-long and to get a long neck is potentially available in the library of DNA. But the actual control system is in the RNA and its smaller brothers m-RNA and t-RNA. Smart balanced systems within the cell and cell nucleus and communication between cells, tissues, organs and even the outside world ensure that there is "a button" you can turn on or off. A set of genes is turned on, changed or turned off. A long neck and long legs grow in generations to create the giraffe. The mechanism is extremely complex and still not very well understood.

In any case, everything is based on "exformation", the powerful filter in nature that cancels useless information and reinforces essential information.

It is therefore no surprise that a simple broad bean has 100,000 genes on the DNA and a complex being as a human being only 25,000. After all, man has developed so well that it has dumped the useless and unusable information in the process of exformation. The big question, of course, is why all usable and unusable information on the DNA was developed so early in the evolution. According to mathematicians, that has to do with the laws on complexity. As soon as a complex network develops (and that is already existing in a simple cell), complexity exponentially expands to even more extreme complexity. Ultimately, this complexity may behave as an intelligent learning system ("natural intelligence"). I do not speak about intelligent design, because that also suggests a designer, but "intelligent outcome". Gene systems evolve automatically to learning systems that also have a database in the DNA with successful traits.

The useless information that is not used for generations is eventually erased. Even then, the hint of a supernatural super intelligent power has not completely disappeared.
Although I still reject that for the time being, I am convinced that the entire cosmos is a learning and intelligent system. After all, it is only information, there is no real time, matter or space. The cosmos is materialized by consciousness (holo-consciousness) which is capable of transforming a timeless, space-less and immaterial quantum world full of information into the familiar and mostly unknown cosmos that gives us so much wondering. This leading to complex systems as living intelligent beings is no coincidence, but rather a logical consequence. An "intelligent outcome".

Chapter fifteen

The Fermi paradox or the paradigm?

The American physicist Fermi (born in Rome in 1901) has suggested that according to statistics there must also be intelligent life outside our planet elsewhere in the cosmos, but that it is a paradox that we haven't observed it. This statement is called the Fermi paradox.

In the scientific world there is no evidence for the existence of E.T. so, the Fermi paradox still stands. Actually, given the size of the universe and the recent evidence of the existence of exo-planets, you can't either intuitively or rationally, suppress the presumption that we are not alone.
Why didn't we discover aliens then?

There is no simple scientific answer to this, but we will have to be philosophical.
First, we must realize that our equipment that can explore the universe is still very young.
Living in space means that we always look to the past. After all, the speed of light is one million km per 3 seconds (550.000 miles/3 sec), but the universe is so empty that even the nearest star (Proxima Centauri) can only receive a signal from us after almost 4 years. The stars in the night sky have sent their light to us millions or billions of years ago and most of them no longer exist. Conversely, aliens see us in our past, a planet 1000 light- years away from us, sees us walking around in the Middle Ages.

Suppose you live in a metropolis where messages are sent by small snails that can travel 1 cm per day. No other means of communication is available. You send a letter to your uncle who lives on the other side of the big city, 30 km away. How long does it take for the small snail to deliver the message to your uncle? That's 8219 years 2 months and 4 days. By the time the mini snail delivers the letter to your uncle, you and your uncle have long since died. This shows how pointless communication in the universe is, because the situation is that the mini snail in the real universe is even actually much slower than in my thought experiment.

Drake, a mathematician, made an equation that calculated the probability of intelligent life in the universe with the data they had at that time. I will not discuss the details, but it was clear that this chance would be very small. So, we live in a metropolis where most of the houses are uninhabited and only a few streets are home to people who can send mini-snails on their way. It could be that our planet is the only one with intelligent life in the entire Milky Way (100 billion stars) A weakness of Drake's calculation is that in the model "earthly intelligent life" was used. The assumption that the conditions prevailing on our planet are the only possibility for the emergence of intelligent life elsewhere is very speculative and will probably be an underestimation of reality.

Meanwhile, it's also been scientifically proven that the cosmos is teeming with exo-planets and the chance of life elsewhere has increased considerably.

It's questionable whether every planet with life by definition also develops intelligent life. When we look at our own planet, evolution has taken almost a billion years to produce a creature with a complex brain that makes intelligence possible. These creatures (mankind) are evolutionary alive only 1 million years (one thousandth of the total evolution time) of which only a few thousand years in the civilized form, a few hundred years in a highly technological form and only half a century on search for extra-terrestrial life (SETI) *).

Our mini-snail is just over the door mat and not even out of the front yard on its way to our uncle!
The big question, of course, is whether we survive as a species long enough to get back the messages from "our uncle". But also does "our uncle" live long enough?

Finally, there is the "mini snail problem". Does our uncle know that these mini snails are delivering our letters and can he read them too? Can our uncle send a letter back in the same language and will the mini-snail know the way back in a city of more than 8200 years old? After all, the city has been nicely rebuilt after so much time. Is our house still there after 8200 years? Do we recognize the mini snail of 16400 years old of our uncle?

In parallel to the real situation, it may also be that we are one of the first intelligent civilizations in the entire universe and that we may not survive long enough to discover whether even more equivalent civilizations will arise. The reverse is also possible, but slightly less likely. If we are the last intelligent civilization, the chance that older highly developed civilizations elsewhere have left their traces from a distant past is much greater. After all, we always look back to ancient times. To make matters even more complex, the universe is also expanding rapidly and the distances will eventually become increasingly larger. As a result, our mini-snail has to travel ever larger distances in an exponentially growing city. Eventually the observable universe will continue to thin out and

less and less stars will be visible to the night sky because the light of this can no longer reach us.
Fortunately, the expansion is relatively slow.

An uncertainty is that we do not know if the house of our neighbours (that our mini snail can reach after 4 years) may have been inhabited. But if so, SETI could have already discovered it 46 years ago, but again: if there is a civilization, are they in the Middle Ages, prehistoric times or already in a much higher technological phase? That's uncertain. And would they be able to receive our electromagnetic signals or do they have developed very different methods of communication? For example, communication via gravitational waves which are difficult to detect?
In short: the window of opportunity to communicate through the interstellar space is extremely small and probably nil.

Fermi suggested another possibility, namely that it may also be that E.T does not want any contact at all and that this is the reason for the Fermi paradox. Why would they want to contact a civilization that is potentially aggressive? But I think that if they are as curious as we are (and that is actually a characteristic of intelligence) they will certainly have the need to communicate with other civilizations.

My conclusion is, ultimately, that intelligent life is far too short lived, too far away and because of that, all civilizations in the universe have the same fate to think and wonder if they are all alone. Maybe that's the ultimate fate of intelligent life, "a splendid isolation".

The question is, of course, whether that is a blessing or a curse. Maybe it is better to speak of a Fermi paradigm (a scientifically prevailing insight) concerning the fate of intelligent life in the universe.

Oh yes and then again: if you read this story you can understand that it is absolutely impossible that we have ever really been physically visited by living intelligent beings of our own size. The slowness of large massive spaceships is so enormous, because the amount of energy required to give these objects even fractions of the speed of light. This makes such an operation practically impossible, even with advanced technology.

Nano spaceships with artificial intelligence are perhaps an unavoidable alternative. Their mass is so small that energy management plays no role and because of their invisibility to the naked eye they will leave no trace for the visited civilization.

It is therefore exciting to learn that a Russian billionaire (Yuri Milner), promoted by the late physicist (Stephen Hawking), is launching an initiative to send nano-ships to Alpha Centauri with 20% of the speed of light. Estimated time of arrival 20 years after take-off. First message will reach us 4 years later.

In conclusion, there is no reason to suppose that if we see an UFO that we are dealing with aliens, on the contrary, we can immediately scratch that option.

The paradigm is that aliens will never be able to visit us.

*) SETI = Search for Extra Terrestrial Intelligence

Chapter sixteen

Consciousness and holo-consciousness

Has the universe created consciousness or did consciousness create the universe?

Can there be a universe if there's no consciousness? Can there be a universe even if nobody perceives it? Can quantum collapse occur without consciousness? (Spontaneous quantum collapse)

These great philosophical questions could be answered right away when cosmologists were tomorrow discovering that consciousness waves existed.

Suppose they existed, what physical unit would these waves have? At least a measure of information, something like bits, something comparable to a quantum, but not with an uncertainty factor, but a certainty factor. All very speculative, but let's play with it.

However, it seems that physics and the human sciences can hardly be reconciled and that is not so much related to the content of the academic education, but more to the existing presumptions. The physicist thinks reductionistically and will eliminate any disturbance by human interpretations.

This branch of science uses basically mathematics and can thus make predictions and descriptions of the reality that at least have a practical outcome. This allows planes to fly, and takes missiles to Mars. The andragogic scientist thinks in fact from human interpretations and will use psychology, sociology and history as the basic sciences.

Thus, consciousness for a physicist is an unworkable concept for which no formula can be conceived, while the andragogic scientist sees consciousness as a self-evident and central phenomenon from which all human actions can be explained. Moreover, no philosopher, neuroscientist or psychologist has ever succeeded in making consciousness into something concrete.
The essential question is how a material system or network, such as our brains, is capable of creating an image that is a representation of the surrounding universe, our world, but which is also capable of presenting it, for example in stories, films, art and books?

For a physicist, Rembrandt's painting "Night Watch" is thus (just exaggerating) a layer of paints with certain optical properties (reductionistic thinking), while the andragogic scientist sees a brilliant imagination of a 17th century Dutch painter who was a master in the representation of atmosphere and light (holistic thinking).

What reality really exists? What perspective is useful in this?

Anyone who reads this, will of course say that the holists are right when it comes to the Night Watch, but that doesn't make the reductionist world view untrue, because otherwise planes wouldn't fly and Einstein and Edison would be charlatans.

Here we come to the essential point. We can't define our consciousness along the reductionistic path (how can a few billion connected dumb neurons in your brain, as a whole, think up something very clever?) Neither the holistic way (can you point out where my "ego" is or where my consciousness settles somewhere in that bony skull? How much does the "soul" actually weigh?).

I therefore propose to think "out of the box" about consciousness. It isn't something that resides within our skull, but it is a universal phenomenon and is not local, just as the quantum world is not local. After all, space and time are only illusions of our brain and our consciousness.

But why is it that experiments with our brain show that there is a connection between the neuronal stimuli somewhere in the cerebral cortex or other parts of the brain and our perception of reality? It is true that during brain surgery certain sensations can be generated by electrostimulation of brain cells. After all, it is a physical stimulus here that arouses consciousness, isn't it?

It is also true that we do not perceive anything without an optical cortex, that drugs can make us psychotic, that disorders of our brain chemistry can make us depressed, that strokes make our consciousness incomplete and dementia can deprive our whole being! You can't possibly deny that consciousness and physical brains are depending on each other.

But actually the discussion is not about that at all. Human consciousness can't exist without a brain, that's is no point of any debate, but that doesn't explain what consciousness is and how the activity of neurons in the cerebral cortex create our representation of reality.

As I said before, I assume that having a conscious experience doesn't initiate in the brain, but comes from an universal consciousness (holo-consciousness).

Back to the consciousness waves as a hypothetical model. The question remains whether human consciousness is an exponent of that universal consciousness, thanks to the presence of functioning brains, which can interact with those hypothetical consciousness waves.

You could compare the holo-consciousness with the Higgs boson that gives mass to particles through the gravitational field of the universe.

Completely parallel to this, the holo-consciousness could give consciousness out of a universal consciousness field to the brain's neurons. That will lead to a local experience, more specifically an experience that will fit with the immediate daily environment of that person at a certain time.

Thanks to our senses, the brain is capable of doing this. The information from the senses comes relatively slowly via the electrochemical pathway, but that isn't a problem, because the local experience is more a continuum of experiences and not a momentary phenomenon.

The "downloaded" consciousness waves would follow the quantum mechanical path and is "real time" (see the section on quantum consciousness) and would be instantaneous.

So, the human brain could only acquire a local consciousness by interaction with hypothetical "consciousness waves" in the universe (phenomenal binding according to philosophical jargon).

This of course also sounds very vague, elusive and daringly speculative, but so were once quantum physics, and yet it's still the best physical description of reality to date.
Dark matter is perceived indirectly, but unapproachable by physics, let alone dark energy, only just to name a few. Yet we know that it exists, just like consciousness.

Consciousness can't also be explained entirely by the five senses. Suppose you are deaf, then you can still see and form an image of the world. If you are deaf-mute you can still form a picture of reality (if you have ever observed it) And even if all sensory perception is taken away from you, you still have thoughts and feelings and you know that you exist. After all, the inner world always exists. No matter how terrifying this vegetative state might be, you still have consciousness without senses, provided you have had those sensory stimuli before.

Finally, I would like to point out the near-death experiences. It can't be explained along the materialistic scientific way how a dying, oxygen deprived brain can produce such vividly clear experiences.

The scientific explanations so far, are very artificial. It is often claimed that lack of oxygen in the brain is the cause, but that seems very unlikely to me. The experiences of a near-death experience are so precise, vivid, detailed and confirmed by thousands of separately examined patients that it is highly unlikely that these are the hallucinations of dying brains. In addition, it is a scientific fact that hallucinations are requiring extremely much energy from the brain and you can't expect that in a low-oxygen environment.

Also, the comparison with the hallucinations of super fit fighter jet pilots, who go hallucinating in case of oxygen deprivation in the cockpit, is completely flawed, after all, the hospital patient is far more defective in terms of blood supply to the brain and also has them. In addition, recent research has shown that up to 15 minutes after death, there may still be brain activity in approximately 20% of the cases studied.

My conclusion is that you have to seek the explanation of near- death experiences in the transition from local consciousness, caused by brain activity, to the non-local consciousness, the holo-consciousness. After all, due to the lack of oxygen, the brain is falling out as a "localiser". So, it is more a transitional phase from one form of consciousness to another.

Finally, death itself. If you die, you would say that all information should disappear from your brain. That has to be this way, because the brain will decay. However, this is not in accordance with the law of conservation of information. Because where stays all that lifetime information stored in the brain?

First of all, it turned out that information in the brain is dynamic and that there is no real storage of it as a whole. It is fragmentedly stored in synaptic clefts, but not active. When we sleep our brain resets the information, temporary stored in the synaptic clefts, and restores it to use it again the next day.
Useless information is deleted or deactivated, the exformation process I mentioned before. So as in the case of a computer, the information is not stored somewhere on a hard disc in the brain.

So you could say that the conscious information of the brain is just as emergent as, for example, the particles in the quantum world. The information is somewhere, but also nowhere as a whole and not somewhere localized in the brain. It is therefore more likely that information that was once part of conscious experiences of the deceased person would be transferred to the non-local world of holo-consciousness. I come back to this in the following argument.

I didn't say, that I think there must be an afterlife, on the contrary, but rather a return to, or a decoupling from the local conscious state to the non-local conscious state. This corresponds completely to the law of conservation of information. At the same time this also corresponds to the disappearance of the person from our world, after all the brain is dead, the localiser is gone.
The person has gone for relatives who are still localized in time.

According to the entropy laws (second thermodynamic law of Bolzmann), information processing costs heat (energy) and information being released would generate heat (energy).

I do not go into detail for this for practical reasons, but this will be discussed later in this book.

You can purely physically speaking say that heat leaves the dead body because the information that disappears from your brain-body system has generated it.
But that doesn't hold if you consider what heat is already released when you turn off your PC. It's obvious that's only a minute fraction of all life time information in your brain, what would be released as heat from a dying human brain. The heat released by the dead body can just be explained as the heat loss that occurs as a result of fall of all molecular activity of all cells.

If really all life time information from the brain were to be released as heat, the body would spontaneously ignite. (A cheap cremation!)

Why is the information about cell activity converted into heat and heat from the information that is released from the brain, as the memory of conscious experiences isn't? It looks like it that information released from the brain when we die, doesn't seem to obey to the known rules of natural laws.

The explanation seems simple: the information from our consciousness has not been present at all in our brain all along our life. It never has been there. It has only been connected with it and that is not the same as containing it. Consciousness is not at all local. It is an integrated part of the non- local holo-consciousness translated by the brain to a local personal consciousness. After death, the brain body system decouples from the holo-consciousness. The person is gone.

But the mystery of the mechanism remains, just like dark energy, dark matter, black and white holes. We can only make hypotheses that approach reality as closely as possible.

Chapter seventeen

Artificial Intelligence yes, but worldrule no!

There is often confusion about artificial intelligence (AI). AI is not the same as artificial consciousness. AI is a learning program, a written algorithm for a computer. The difference with an ordinary algorithm is that the output is not determined by the algorithm alone, but by the result of the learning process of the algorithm itself.

A stupid computer can quickly calculate what the orbit of a satellite will be, but an intelligent computer can predict how, for example, a flock of birds will behave which requires observation and drawing "conclusions" about the behaviour of each individual bird in order to predict how the swarm will behave. The human brain is not capable of such a performance.

Here seemingly decisions are made or answers given (output) that in terms of complexity are reminiscent of human intelligence. A good example is a strategy computer game such as "Civilization" or "The Sims". The computer learns on the basis of pre-programmed rules how the game should be played. AI is also used in science to solve mathematical problems, to make economic simulations, to imitate chaotic processes (climate studies, weather forecast). In the media, science fiction and film industry AI is often given an exaggerated human face. Films such as "Star Wars", "I Robot" and "Space Odyssey 2000" are examples of this. Robots and computers are depicted here as devices with human characteristics, such as humour, vanity, fear, grief or aggression. The fear of mankind that robots will ever take control of humanity originates from it. It was Alan Turing, the famous British mathematician and father of the AI, who invented the Turing test, a test where a person and a computer can ask each other questions and people have to guess whether they are dealing with a computer or a human being. To date, no computer has ever been able to pass the Turing test. Even toddlers surpass the test better then AI does at this very moment. *)

(* As I write this, I understand that in 2014 a robot passed the Turing test).

The combination human + computer is a very powerful one as we know, but especially from our perspective of progress.
Science would have long since met a dead end if we didn't have computers, because calculations that were previously impossible could now be carried out with great ease and tremendous speed and the mind of man is since then capable of sketching a new model of a scientific reality.
Theorists predict that we will reach a point where the computer outsmarts us and we call this "The singularity".
The processing of the computer would then be superior to that of the thinking in human brains. The tricky thing about that prediction is, that we do not precisely know what a thinking process is.
Thinking processes takes place in our mind and is not some calculation, but is characterized by conscious sensations, usually accompanied by inner visualizations, images or language and of course any combination of it.

The philosopher John Searle invented a thought experiment that he called the "Chinese room".
It's described as follows.

In a closed room there are people who receive assignments through a hatchet written in Chinese characters, and although they don't know a word Chinese, the people in the room know exactly what the instructions are and that is why they produce a right outcome through another hatch on the other side of the room. There's always a good answer to the question, that was asked in Chinese. With this thought experiment Searle elegantly shows that computer intelligence is a process and has nothing to do with genuine meaningful understanding.

Sir Roger Penrose, the famous mathematician, showed in his book "The emperor's new mind" that no matter how clever AI becomes, it will never know consciousness. A computer will never have the experiences that a baby, toddler, child and adolescent have in, say 20 years, as a learning process in a physical world full of pain, pleasure, sadness and happiness.
After all, these sensations are essential for achieving genuine self-awareness that makes us what we are: living conscious intelligent beings.
This brings us to the point where we have to ask ourselves what is the difference between our own mind and a robot who, after reaching the singularity, is equipped with modules that can simulate pain, pleasure, sadness and happiness acquired in a learning process that takes place in a humanlike mechanical body.
Such a robot could grow up, for example as a kind of "pet" as part of a family with upgrowing children.
Even then, the robot will gather a lot of input about human behavior, but will have no experience whatsoever.

The robot can still be turned on and off at our will, does not change after a reset or if replaced with certain parts and will have, in theory, an eternal existence.

But even then, it still remains a zombie, whatever good the simulation of human behaviour may seem. It is notinconceivable that we, like our pets, could get a bond with such a robot, but that would never be mutual.

The robot might be able to simulate crying if a family member would be harmed or die, but never experience any real grief about it. The other way around could surely be possible, in fact would be very likely, given our very common response, when, for example a beloved pet dies.

The distinction between man and robot will always be determined by our consciousness, which is apparently something different than even the most complex algorithm of a computer or robot.

Our brains are in fact very flexible, they grow and change every minute and the billions of neurons constantly make new connections with each other. There's no day you wake and go to sleep with the same brain.

Another important detail is that the brain works both biochemically and electrical.

Though these are just technical issues, they show that living brains are incomparable with silicon chips.

In a human mind there's no algorithm at all! Even our thinking happens in an associative, quite chaotic matter, based on routines and previous memories and experiences.

Apart from the mechanistic difference between biology and technology, there's also an essential philosophical difference.

Humans realize early in life that they are mortal and this drives them lifelong. Plans are made, we receive education at school and we're brought up, we start relationships, all just to let the species survive.

This is to ensure that life is a meaningful whole, and ultimately that's also in the interest of our species as a whole.

That's how progress in science, engineering, arts, culture and a general level of civilization is made, making humanity a well-organized surviving machine.

Such an existential drive is completely absent in the "artificial mind" of a robot, after all, it can easily copy himself to another machine and doesn't have to make any effort to let its "species" survive.

A robot will be totally indifferent to its own existence, though it could "exist" eternally on a technical scale. It doesn't live as a biologically built creature with a mortal outcome and has therefore no existential will. My conclusion is that AI is a very powerful tool in the hands of mankind, but will never be equal to the level of thinking and level of consciousness of man even after reaching the so-called singularity.

Our ability to recognize patterns and sense dangers more intuitively will
still make us superior to AI.
Yet there are dangerous developments in the world of AI and it is amazing that so
little attention is paid to this in the media and politics.
The military apparatus is planning to use robots on the battlefield that can decide
to independently kill enemies. This may save friendly soldiers, but can you let an
algorithm decide on life and death even if it's the enemy? Where's the limit,
because who decides who the "good guys" are and who "the bad guys" are?

What if the bad guys can make such a robot? How does it end up for the good
guys? Will North Korea then dominate us with their killer bots?
It looks like science fiction if it was not so frightening up-to-date. World citizens of
all countries join and stop the killer bots!
We certainly are late in realising we have to create international laws now, to
limit the rise of the robots in the very close future.

Artificial Intelligence yes, but worldrule no!

Chapter eighteen

Brains in a Petri dish

W hen I look around me, I am aware.

If I do not sleep, my brain works normally. When I drive a car, I do it automatically, completely unconsciously, but when I ponder a bit during that activity, I do so consciously. There is no one-to-one relationship between consciousness and brain activity. In other words, my consciousness isn't off when my brain is off and vice versa. Moreover, my brain influences my consciousness. My machine, my brain can be defective, poisoned by alcohol and drugs, damaged by a disease or some genetic defect. I will have consciousness, but it's changed by this machine, my brain.

When I'm in anesthesia I hear everything, but I remain unconscious. The ratio between the amount of information in the unconscious and the conscious is one to a million. This means that you keep 999,999 stimuli out of your consciousness every second.

And that's a good thing! It would be impossible to focus on a million stimuli more then you're managing now. It would be absurd to consciously think of how-to breath in and out.

Another interesting point is the OBE and the NDE. Respectively the out of body experience and the near-death experience. In both situations, the state of consciousness can't be explained scientifically by the according brain activity, because it is virtually eliminated. Another special condition is the anxiety and compulsion disorder. Here the brain activity is going to do something that's unwanted for the person and is experienced as ego-strange. Who is the captain, your ego or your brain? I have discussed this earlier. The people who experience it find that the person can usually impose his will on his sick brain, but during intense symptoms the brain temporarily wins and the person suffers very seriously. Mostly so severe that medication is needed to supress the brain.

In a psychosis, the brain is the captain, the person is possessed by his sick brain. In mood disorders such as depression or mania, a similar struggle occurs between the person and his brain. In dementia, the tools that the person needs to function as a person are defective: memory, intellect, character traits, inhibitions, executive functions and planning. The person is conscious, but lives in the present and is deprived of his normal daily life.

If the connections between left and right brain have been surgically cleft, one half does not know what the other half is doing, but you remain a conscious person. The brain can even be destroyed for 90%, with the person still being

ꞇ

aware and functioning reasonably well. Experimental animals, having been stripped of the auditory center and changed by visually functioning neurons, effortlessly switch to hearing with visual neurons and seeing with auditory neurons. Retransplantation gave the opposite effect. This means that there are no specific neurons for hearing or seeing.
Patients with a stroke (cerebral infarction) miss a substantial part of their brain, but remain albeit handicapped) conscious persons. In short, from all these facts listed, you can draw the conclusion that brains are indispensable for maintaining consciousness, but that consciousness is not just brain function.

Let's take a closer look at the process of seeing.
A ray of light hits the retina and strikes a light-sensitive cell that generates an electrical signal in the optic nerve. Through a centre in the medulla oblongata and some relay stations, the optical cortex (a field of millions of neurons) ultimately creates the conscious sensation of seeing, for example, a rose. The problem is: Is this sensation of seeing a rose just the electrical activity of a group of neurons in the optical cortex and does it stop there?

Let's suppose that there are neurons we call "rose" neurons, because they only recognize roses and can create the sensation of seeing roses. If we turned off these neurons, the person in question would never recognize roses again. The transplants in the brains of experimental animals clearly show that neurons are multitasking, so no specific "rose neurons" can exist. But where does the "rose experience" generate then? Do conscious and unconscious neurons exist? There's no proof of this at all, on the contrary.
Descartes thought that the pineal gland contained consciousness. Even now people are looking for such a place in the brain. It will never be found because it isn't there.

Reductionistic science can therefore not deal with this problem and holistic reasoning can. The sum of activities doesn't make the whole, the whole exceeds the sum. It creates a new state. Philosophers like Daniel Dennett do deny the existence of consciousness just to escape from the problem.

This shows how childish science deals with phenomena that are so elusive that they can't be studied. I can accept saying consciousness is "a mystery" but not a non-existent feature of the brain.
What is certain about consciousness? Actually, nothing is, really! We suspect that brains are necessary to maintain consciousness, because dead brains are no longer conscious.
The deceased in question disappeared as a person after his brain died.

At the moment, experiments are going on worldwide with stem cells. These are primordial cells similar to the fertilized egg that are everywhere in your body and are only awakened when a tissue or organ is damaged or needs replacement. This process is called regeneration.

For example, your bowel and skin are replaced several times a day, skeleton a few times in your life and your brain never.

Yet it also appears that the brain can partially regenerate. For example, doctors have succeeded in treating Parkinson's disease with stem cells by injecting them into the damaged brain area (the substantia nigra). If we know how we can program stem cells, we can also grow organs in a laboratory that serve as a replacement for a diseased organ. This cultivation outside the body is called an in vitro culture, in the body in vivo. Because the body is the only place where stem cells get detailed instructions about building plans, in vitro cultivation is only partially possible. You can grow tissue with liver cells, but not a whole liver in the Petri dish (breeding tray).

Would it also apply to brain cells? To a limited extent, as in the case of specific diseases of the brain, but it is certainly not inconceivable that in the distant future Alzheimer's disease can be treated in such a way and perhaps MS, schizophrenia, depression, bipolar disorder, anxiety and compulsion disorder, autism, ADHD and psychopathy. A psychiatrist will inject some stem cells in you to cure you!

The possibility of replacing the complete brain might become possible in a few centuries. Imagine agreeing with a friend in advance that if he gets into an irreversible coma, he will have stem cells from you injected into his brain and that all his connections and neurons will thus be rebuilt to yours.

This creates an exact anatomical copy of your brain in his skull. If for some reason, at any given moment you would almost die from some physical disease or from old age, the copy of your brain in the skull of your comatose friend (your back-up brain) could be transplanted to your skull and body. Of course, your friend dies, but you just continue to live, thanks to your back-up brain, provided that the physical problem is also sufficiently solved with stem cell therapy to provide the brain with sufficient nutrition and oxygen. But who will you become?

1- *you get the consciousness of your friend*

2- *you live on with your own consciousness*

3- *you become a zombie*

4- *It is impossible and your consciousness disappears, you die.*

ͼ

The first option is that you would get the consciousness of your friend.
Assuming that the memories and experiences of your brain are copied, this option
seems very unlikely. After all, you are then reminded that you lay dying and was
awaiting your backup brain. If you have failed to copy the data of yourself into
your new brain your set up failed completely. Your friend however would not be
dead, he just lives by awakening from his coma. So, you die after the
transplantation. Yet, it is then your brain and it is questionable whether the
consciousness of your friend would feel at home in another brain and another
body. In short, there are too many inconsistencies and obstacles to the ego
foreign body where my friend should live in. Practical problems would be so
great, that it would be impossible to live a normal life. As my friend, for
example, was left-handed and I right-handed, the problems already start about
minor issues. But, as said, that's just a small thing compared to the other
interaction between my body and his mind.

The second option is much more likely than the first.
Indeed, you are a repaired copy of your brain with your own body. Conscious
experiences are momentous and will start up the minute you wake up from the
surgery. Wouldn't you change in a new hybrid person with a strange brain in
upon your body.? After all your "persona" is never the same, not even in the
normal situation.

Indeed, fraternal twins are exact copies of each other, but they each have their
own consciousness, not a common consciousness.
So I think you will never be the same person you will feel related to some
strange memory of your "one fraternal twin." Even if your memories are
preserved these memories will be from a "previous life" and therefore not
belonging to the ego.

The same problems would arise having another body, the body of your friend.
Supposing the transplant wasn't necessary to wake the body of your friend up.
What about the relatives and beloved ones of your friend, would they be hostile to
you as a body snatcher? Is your friend left or right handed? And these are just the
minor problems…!
We have to remember that consciousness is a state we're in and has in fact no
material substance in a brain.

The third option is that you become a zombie, with your "persona" in it.
In that case, you have no more consciousness. It really comes down to dying.
However, your brain body system functions quite normally (as a living robot)
and no one notices that you have become a zombie, after all, human interactions
and behavior will be executed and simulated perfectly by your brain, but your
consciousness is gone, you will have no experience at all. The drama is that your

1

relatives and beloved ones still think your ego is in you, while nothing of your mind is left.

The fourth option is that both your friend and you die both after the transplantation. This is similar to option three, but goes much further. The transplanted brain will not survive in your guest body and therefore you die not only as a person, but also physically.
Apparently then, your consciousness is unique and non- copiable, but is probably also essential to keep your brain body system alive. After the brain transplantation there will be no backup brain activity in the new brain due to the lack of consciousness. We know in neurology that people can be comatose for many years and we also know that in rare cases people recover from that vegative state and function normally afterwards, though have no memory at all about those years in coma. This can be a memory loss, but it's more likely that there was no consciousness at all in that comatose period, because brain activity, being measured wit EEG was minimized in that period.
So, consciousness is not the main reason that you stay alive, but life is at least necessary to have consciousness. If a brain is non copiable to another body , which could certainly be possible, then this would be caused by the fact that a lot of otherfunctions that are regulating body functions are disturbed because your body and the body of your friend are very different.
In all cases your goal to live on would come at an unacceptable price, becoming a zombie, extremely unhappy or you would still die.

Have you already chosen what you find the most likely option? Of course we will never know it for now, but for us understanding consciousness this thought experiment is essential.

In my book I explain the driving force of the universe is consciousness. The universe has a consciousness of its own and I call that holo-consciousness. That's because the cosmos is in fact not made up of matter, but of information. Information needs an information system which is holo-consciousness.

To manage all information about the cosmos, holo- consciousness must require superior processing. As humans we think the only species with mind, intellect and memory are we, ourselves. Or our brain. We experience a material world and so we interpret all we observe in a materially way.
If you replace matter with information, the entire cosmos is a completely abstract world. We can't mentally deal with that. So, our brain will always translate that into a simulated material world in which we feel comfortable.
An important indication that our world is not quite material is explained by quantum physics.
This stresses again the importance of an interface between quantum reality and simulated reality, which is our brain.
Copying brains therefore makes no sense. Information is immaterial, so you can't copy experiences and memories out of a brain, because it's not there. It's non-local.

Indeed, the brain is the only material substrate of consciousness.

The brain can't be meaningful without consciousness. In fact, doctors pronounce you dead when a brain doesn't experience any consciousness anymore.
You're not a brain, but a conscious state in an "apparently" material world. That's why consciousness simulates being unique in place, time and person and your brain is doing that for you.

But in reality you're a child of the universe, an integral part of holo-consciousness and hence irreplaceable.
Your being is carefully precalculated for and functional for the universe such as chaos theory suggests.
 Brains exist to navigate in a seemingly material world, which in reality is a quantum world, being collapsed by conscious observation .Holo-consciousness creates the seemingly material universe and not vice versa.

That, dear readers, is a radical idea, but explains why we find no centre of consciousness in our brain. Consciousness is non- local.
Now I will also frustrate all people with religious or spiritual beliefs severely
 Holo-consciousness is not to be believed at all transcendent or spiritual. It is just the ultimate basic reality of everything and the consequence of a universe consisting only of information and not of matter.
Science should eventually be capable to understand this reality and not some holy mythological book or superstition.

Brain research over the past half century has really taken off. Brain research is like investigating a complicated engine through a keyhole in the dark with a burning matchstick.
I have told you about the brain research of Lebit before, which was questioning free will.
But even if you accept that consciousness has a purely passive role, it hasn't been proven to be insignificant and that's where neuroscientists make the error naming it an epiphenomenon.
Perhaps the existence of consciousness is just a goal of nature itself and is the brain merely a good machine for it. Scientifically we can't answer this question, nobody can.
An argument of evolution biologists is that consciousness is mainly necessary to stimulate our survival instinct, because we recognize a totally irreversible loss of consciousness as our death, which we instinctively fear. We just want to postpone this frightening situation as long as possible.
But then, dear readers, the function of consciousness is still unexplained, it can easily be seen that animals with minimal conscious states and even plants or trees make every effort to survive without consciousness or awareness.

1

In modern times many neurological studies use brain scans. The scan can show excellently where the activity takes place in the brain when performing tasks, thinking and observing. It certainly gave insight to the physiology and functioning of our brain, our machine, but nothing about consciousness. And where Lebit doubted, today's researchers should also ask questions.

Can you really visualize a conscious sensation in a fMRI scan of the brain? Here we enter the world of neurosceptiscism. They really put it in doubt whether consciousness is only a matter of a few billion electric currents somewhere in a cauliflower-like complex biological organ (the brain).

Instead they suppose it is seemingly only experienced there, but does not necessarily originate from it.

Alva Noë, philosopher, neuroscientist and neurosceptic wrote in his book, "We're not a brain" that consciousness can at its best be seen as a stream of experiences in a certain timely order within a discrete subjective time span and that we cannot grasp the concept itself, because we ourselves are part of it. He claims that the brain can't think itself, but rather creates the conditions for us having conscious thoughts. He points out that most of the information in our consciousness comes from the outside world and that someone who has no senses to observe the world can never reach a stage of consciousness. Implicitly he says that consciousness is in fact an information stream flowing into a receiver that makes a movie from it. It is totally dependent on the outside world and can't itself be ignited only within the central nervous system in our skull.

Let's go back to the idea that our brain is also a quantum computer that has mighty properties. I can't explain quantum physics here, but just remember that time and space do not really exist in the quantum world, but is purely abstract information (a number of bits) that has a certain probability to be somewhere in the cosmos in place Y with speed X. The bizarre result of these laws of nature is that particles can simultaneously be in two places at the same time and can transport information without the use of the speed of light to any place in the universe.

Even more fascinating is that a conscious observer eliminates all these quantum laws when he sees the particles. Simply put, every particle of light you receive with your eyes is coming from the quantum world and only by conscious observation then becomes part of the everyday world that you look at. So, the rose in the previous text, is only a rose as a result of the collapse of the quantum properties of the rose by just looking at it. If Sir Roger Penrose is right, we observe the world, in fact, by means of a quantum measurement system. Roses, cars, people and houses, etc. The optical system in our brain primarily functions to verify this as a check-up of a collapsed quantum state to a reality as we know it and pin down this information. This process works with a much slower chemical neurological process and therefore only as a backup for our momentous "quantum" consciousness.

Consciousness can therefore , given the arguments I brought here to you, not purely be a side effect of brain activity as we just put these things straight.
It may even be an overall cosmic phenomenon that in fact holds the foundation of a perfect, irreproducible and unique universe, which we briefly share as a person.
The theistic view, that we actually walk around in the virtual world of a thinking God, is the foundation of Spinoza's famous book "the Ethica".

Postscript: I chose option 4 to be most likely.

Chapter nineteen

What makes nature invent itself?

Humanity itself is hardly capable of inventions, even if we invent aircraft, missiles, particle accelerators and nuclear bombs, all fundamental true inventions come from nature, and we just copied them. Maybe you would mention the wheel, but I'm convinced primitive humans got inspired by the circular shape of the moon. The patent holder and engineer is mother nature. Almost all great inventions find their concept in nature. The number of examples is almost endless, but I will not mention them all. A few we can't leave undiscussed. The wing, the eye and the nano-engine.

The Wright brothers invented the airplane. They understood that a wing, by its structure, develops lift in a laminar airflow. We know the follow up, but aviation only exists 100 years and all flying animals a staggering 800 million years. Insects are one of the most successful flying creatures of evolution and never got extinct. They were followed up by the flying saurus (pterodactyls) and birds.

Louis Daguerre is considered as the inventor of the first workable camera, but the camera obscura was already known in China when we were still barbarians running around in bearskins.

The eye is 800 million years old and they were probably already working in the earliest Cambrian creatures.

Although the principle is simple: a cavity, with a hole in it, catches a diffracted beam of light which is projected onto the back of the cavity. This creates an image of the outside world. But it is a mystery how nature even "knew" that there is light at all in an unknown outside world. How could it know that light could be used to observe the environment and navigate through it?

The latest invention is the nano motor for which my compatriot Feringa received the Nobel Prize in chemistry. These mini motors are only a few molecules wide and no more than one millionth of a millimeter. But they already existed in the first cells 800 million years ago. It is unimaginable that the most primitive unicellular organisms already possessed these molecular motors that can run, roll, slide, run, shrink, expand and sway.

The science that studies the technical achievements of nature of is called biomimetics.

As we get to know more and more about the achievements of nature at the nano-level, we can develop nano technology and will continue to make complex machines that are smaller and smaller. The model of natural selection by Darwin as the engine of evolution is no longer plausible as we observe these improbable complex structures already 800 million years ago. In the book "At home in the

universe" (The laws of self- organization) by Stuart Kaufmann, the author presents a mathematical model which describes the development of useful complexity. Unfortunately, he doesn't explain why and how nature is capable of self-organization. These might be natural laws that transcend the ordinary laws of nature and are even supernatural?
I do not think so, but we need to look differently at intelligence. Intelligence is needed to develop a simple structure into a more efficient complex structure, which is then developed further again. The most intelligent beings we know are we, ourselves, humanity with the most complex organ we know, our brain. We see our brain as the source of our intellect. It is therefore not surprising that we suppose a supernatural being with infinite knowledge and intellect, called God, as the creator of the natural wonders.
But as I stated in my previous blogs, it is very doubtful whether consciousness and mind resides in our brain, located in time and space. The brain doesn't think or has experiences, but is probably fed by thoughts from a much greater cosmic consciousness system. Our brain gives us rather the perfect illusion that we are a person in space and time.

Quantum physics teaches us that something can only exist in space and time when observed by a conscious system. Our reality is therefore in fact created by ourselves and therefore can't exist without our awareness. But 800 million years ago, there were no sentient beings yet. It seems impossible that simple primitive life forms could develop themselves as ingenious as they have done apparently without any intelligent intervention from outside.

For example, the presence of light, and the development of the eye. The word "outside" suggests a divine intervention, but that's not what I mean. There was no divine intervention in nature.
Consciousness is in fact a cosmic phenomenon and has an infinite intelligence. It is just a fundamental property of the universe. Here, also a holistic model is more useful than a reductionistic view. The totality of the cosmos creates a more superior whole. This, somehow strives for total perfection. Holo-consciousness therefore acts as a quantum phenomenon which ignores the speed of light and the observable universe. It's everywhere and is the eternal catalyst for everything that is in existence.

1

Chapter twenty

Science and religion

Faith is the natural product of the human mind and largely determined the history of mankind. There is plenty of archaeological evidence that religion was already present in the early days of mankind. Initially to explain the mysteries of nature, such as natural phenomena and laws of nature. Even religion has evolved in the course of history. Initially religion only focused on nature and man had the illusion to get control over nature by making sacrifices to the gods. In the civilized world polytheism peaks in the Roman Empire, but with the rise of Christianity and later Islam the belief in one God, monotheism, took hold. Also, in the east faith undergoes changes. The ancient Vedic scriptures are the oldest written documents that have religious content, which later developed into the current Hinduism and later Buddhism. Although this religion worships several gods. It is also called monotheistic, because the image of one god appears in different gods. An important feature of religion is the belief in a supreme being who is omniscient and timeless. With the rise of the Renaissance and rationalism God disappears more and more into the background. Galileo is considered the most famous victim of the clerical authority in the scientific world.

Atheists like to propagate him as the first atheist, but that's a misrepresentation. The issue surrounding Galileo was much more sophisticated. The church was also very engaged in science and there was no doubt in these circles about the spherical shape of our planet. The condemnation of Galileo was merely a political matter not motivated by a narrow-minded religious mindset, but only by corruption. Galileo was also a religious person, not an atheist at all.

Atheism rises in the following centuries paralleled by the explosion of scientific knowledge, especially in Europe and the so-called Western world. Nevertheless, still, 80% of the world population is religious.

However, there is a positive relationship between wealth, education and atheism. The relationship between science and religion has always be problematic. Moreover, scientists and highly educated people are more likely to be atheistic. The tension between the largely atheistic scientific world and the religious world is concentrated in the flanks of the orthodoxy on both sides. On the atheistic side we see a Richard Dawkins (geneticist), Daniel Denett (philosopher), Jean Paul Sartre (philosopher) and in the religious world the conservative side of the Catholic Church and orthodox salaphistic Islam. In the eastern world this struggle is absent because science doesn't conflict with any holy scripture.

By contrast, religious scientists have made great achievements in physics and philosophy, thinking of Einstein, Newton and Spinoza.

Atheism is also not a guarantee for great scientific achievements.

Conversely, religion is not necessarily an obstacle to scientific thinking, for example, the thinking of Thomas of Aquinas.

Orthodox Islam was far ahead to our modern Western science in our dark ages, especially in sciences like mathematics, physics and astronomy.

The tension between science and faith is more of a propagandistic nature. Unfortunately, this affects especially science.

Excluding the existence of God is very unscientific, after all, science should research any possibility, how unlikely it may be if any other explanation can be excluded.

The problem with God or consciousness is that nobody can define its nature a priori. Perhaps humanity will once grow to meta religion which no longer demands us to make sacrifices, to pray, to believe in superstition, handle with certain rites and gather in temples, churches or mosques , but more in the acceptance of an all-overall awareness that nature is what it is because it is a perfect machine in a uniquely perfected universe.

It is mainly Spinoza's merit that he realized that God is just the same as "mother nature."

Religions have one thing in common which is the belief that God is conscious. Most radical in this aspect are the Hindus and Buddhists, who imagine Brahma as the supreme state of consciousness which can be reached by reincarnating.

The so called Abrahamitic religions (from the patriarch Abraham) are believing in a supreme conscious being.

The Abrahamitic religions are Christianity and Islam. Christians really think God is a person, but Muslims don't imagine God at all. At least certainly not as a person.

The relationship between the meaning of consciousness and religion is very clear. There are no religions which worship purely material objects (e.g. a stone) without them having a consciousness at least. The Mayans worshiped the sun, but also found it conscious. Pre-schoolers and toddlers find consciousness in all objects they meet and explore.

This is not a fantasy as we think as adults, but something like a doll or a toy car are really creatures with consciousness for them, because this is the best method for them to train their brain.

There is no distinction between dreams and reality for the small child.

Thanks to the vast imagination of our children's mind it is able to develop intelligence and a memory that surpasses the adult ones by far. The learning curve of a small child is in fact unprecedented steep compared to that of the adult. The pace of information processing in a child will never again be reached in adult life. Research has irrefutably proven that the learning of a small child is accompanied by a massive build-up of the brain.

Once built, this network continues throughout adult life as an unconscious part of the mind, but, as we know from psychiatry, psychology and hypnosis this part of our mind is fundamental for our consciousness, personality and "free will" in adult life. Our brain is essentially a machine that generates a certain state of imagination of reality, which is not reality itself though.

1

In other words, we think in pictures and not abstract, as in mathematics. Einstein also said, "Without imagination, intelligence is worthless." And he could know by his brilliant thought experiments.

Spirituality and belief in souls is a known worldwide phenomenon that has such power of imagination that people can even be "possessed" by it, get strong religious experiences or become entranced just by this power of imagination. Some people can even disable strong sensory experiences such as pain or have vivid hallucinations. Brain research with fMRI shows that the activity of the brain activates many abnormal cortical areas, similar to those occurring in psychosis or LSD trip. Which explains the exhaustion seen after a trance.
Another brain study showed there is a specific brain region, called frontal gyrus, where "God seems to live in." This area is very active in religious or spiritual experiences, even in atheists (!). Moreover, it is proven that rational minded people, (as scientists gladly like to present themselves) are very irrational when they do psychological tests that should prove rationality.

The usual scientific explanation for a God center in the brain is the evolutionary development to perpetuate the social connections within human communities. Like the theory of evolution itself, an unproven assumption and again proving the irrational tendency in science. The Darwinist theory of evolution keeps being disputed.
Believing in God is therefore not irrational, but a natural necessity for the mind to gain control over our dangerous environment.
Still it's a mystery why we have to believe in supreme beings in order to understand the same wonderful world. Maybe this protects us from disintegration, just like "little prayer" when we suddenly find ourselves face to face with death. Yet, most scientists are usually atheists and sometimes even activists, like Dawkins or Denett.
It falls mainly to philosophy to understand the relationship between some belief in a supreme consciousness being, like that of a God or a holo-consciousness, as, for example, defined in the more modern pantheist vision of Spinoza. Unfortunately, philosophy has become a bit of a dirty word at a time of particle physics, Mars rovers and lunar landings. Nevertheless, we will need our childlike imagination and philosophy to unravel reality. Hence the frontal gyrus!

The point is that many atheists deny God for scientific reasons though a lot of science points more in the direction there is one. (For example: the big bang, the anthropically universe, frontal gyrus in the brain and the intelligent outcome of evolution).
The other point is that believers think science is a way to understand how God orchestrated the universe. A lot of atheists hate religion, but are still not certain about their declared atheism when you interrogate them more deeply. The Dawkins test (see end of page) is false because the definition of what God is, is the variable factor in this line of questioning.

1

So, this says, Einstein was a weak theist. In the Dawkins test 3. He just had a different view of the nature of God.

I think, atheists have a more arrogant attitude than believers. Believers are just the result of their education, culture and mental development, this is not arrogance, but maybe ignorance. Atheist just *claim* to know, without scientific proof, that there is no God. This is arrogance or just a new religion. Dawkinism.

Richard Dawkins' Belief Scale Scoring Rubric (from "The God delusion)

- **Strong Theist**: I do not question the existence of God, I KNOW he exists.
- **De-facto Theist**: I cannot know for certain but I strongly believe in God and I live my life on the assumption that he is there.
- **Weak Theist**: I am very uncertain, but I am inclined to believe in God.
- **Pure Agnostic**: God's existence and non-existence are exactly equiprobable.
- **Weak Atheist**: I do not know whether God exists but I'm inclined to be sceptical.
- **De-facto Atheist**: I cannot know for certain but I think God is very improbable and I live my life under the assumption that he is not there.
- **Strong Atheist**: I am 100% sure that there is no God.

Chapter twenty one

Virtual consciousness

As a GP in my office I have a waiting room for my patients of course, but it isn't a normal one. My patients know, because they know I'm researching consciousness. So, first they have to enter the so called "transition zone", which is a room where they have to change their outfit. They must wear a special suite of transparent rubber which contains over a million small sensors and pressure devices. They also have to wear a virtual reality device and a headset. After that start-up they can enter the waiting room, which looks completely normal with chairs, tables, magazines, etc. There's one difference, that is, that behind the ceiling and walls there is an extensive radar system which is scanning every inch of the waiting room. Cameras are also merely everywhere. My consulting room is also completely normal with my desk, bookcase, chairs for my patients, a small examination room where I can examine my patients with all the normal medical utilities. The radar and cameras are also scanning this space of course. Now I, myself, am also wearing the same suite, VR glasses and headset.
There's a very powerful computer in my cellar that is connected with the suits, the VR glasses, the headset en the scanning devices. It's getting all the information of my workspace and all the sensations from the sensory devices from my patients and myself.

So how are my patients doing in this environment? They are doing fine, they don't notice they're in a virtual reality situation. The sound, vision and touch devices are so sophisticated that everybody is experiencing this as really happening to them and they don't realise they're in some unnatural cyber environment. Any bystander without a VR suite looking at like this is seeing an idiotic scene of people walking around with VR devices, headsets and transparent rubber suits. Of course, I'm making exceptions for medical urgencies, they don't have to pass the transition zone.

Now what's the sense of this whole experimental set-up? First off all, I want to know what my patients areexperiencing. After an intensive working day, I open the logbook in my computer and look up the recorded data of a particularpatient and start being somebody else. I pick an interesting patient and put on my suit, my VR glasses and headset. The data from my computer are now replaying the physical experiences of the patient from the moment he or she entered the waiting room and left via the transition room. The data stream to my suit and other devices as recorded that day.

So I don't have to go out there to the waiting room and can just stay home, because all the sensory input is given by the suit, the auditory input by the
1

headset and visual input by my VR glasses. Now I feel myself sitting in the waiting room in a strange body with breasts doing things which are not desired by me, but completely under control of "her will", which actually is only the information the computer has stored.

I'm completely out of control, passively having experiences, because "she" has taken over all my bodily functions. At a certain moment the doctor is calling me and I'm walking will-less into the consulting room.

I see myself talking to myself (the doctor) in this strange woman's body. I'm talking to the doctor (myself) and see myself reacting and questioning this patient and I know exactly what my next text will be, because I remember what I asked her today at the office. I'm very surprised the body that I inhabit does not react already on the next question or is not understanding the doctor adequately.

I did make myself clear today with this question, did I? Apparently not! After the interrogation, I have to go to the examination room and undress my upper body, because I told him I felt a lump in my left breast. When my doctor (the real me) is examining me, I don't feel he's touching my breasts, just because I don't know how a women breast feels. My brain doesn't know this sensation and ignores it. I see him touching this strange senseless breast without feeling him touching it. It seems to be the breast of another person. I also don't feel stress, tension or fear. I also know the rest of the scenario, what the doctor is going to say, what he's thinking and what's going to happen next. I dress myself again and will have to make a new appointment for a mammography.

I'm not worrying about having breast cancer and all the consequences of that diagnosis. I'm completely indifferent about that fact, which I shouldn't be. I should have been shocked about this possibility and wondering how long I will live on and what I should tell my dear husband and children.

Now of course this was just a thought experiment. I'm happy this monster machine is not working in my medical practice for real! What is my point?
Is consciousness something that is interchangeable? Apparently not, you would say. But why? I cannot really place myself in somebody else's body.
Yes, in a way I would feel the bodily responses and sensory input, but would not recognize it as my own or acting in my name.
I still would be me not somebody else.

Now what if I do the same experiment, but put on my own suit and VR glasses and meet the same patient from my own perspective that office hour? Would I have the same thoughts and experiences knowing that I already know what is going to happen next? Seeing the poor worried woman taking a chair and telling me she feels a lump in her breast? Would I consider another scenario, choice of words, formulations or acts? After all, I have already watched myself from the perspective of the woman patient wondering if my doctor is paying enough attention to my unspoken fears.

Though I could consider correcting my behavior towards her, I would still be unable to do so because the computer does not allow me to. I'm forced in exactly the same role and act completely compulsory. Would I resist against myself and try to correct myself? So, which "me" is going to win over the other "me"? How

many rebellious "me's" would I have if I'm playing this game just over and over? How many corrections would I make over and over and how would the patient respond to all those corrections?
So suppose I could change the scenario in the software of the computer? Would the computer know how the woman patient reacts on the changes made in the behavior of the doctor? I'm sure you agree with me it can't, because it's not conscious.

Now wait a minute, I'm conscious, so I can get into the VR world of the woman again and react to the new scenario of the doctor! After having undergone this experience again I would rewrite the woman's role in the software and start over with the doctor's role again and over and over until the scenario is perfect.

Perfect by whose standards?
My standards! Because me having consciousness doesn't make me think and feel like that particular worried woman patient. I can have some empathy or some clue how she's feeling, but it would only be a shallow reflection of that.
The only way to make the scenario realistic is to ask the selected patient to help me and do the scene again, us being us, in our own VR suits. This would be difficult, because we both anticipate on the next event because of foreknowledge, but still we could both write a new scenario over and over again to reach a perfect one we both agree on. But would we both be happy after all that?

So, I showed how it is to be someone else and why that's impossible. I used the method of a thought experiment. Even with advanced equipment it is absolutely impossible.
Crucial is the fact that computers don't have consciousness and can't bring self-conscious sensations to another person's ego even if the simulation is perfect.
Now, I will discuss the relation between time, space and consciousness.

In my VR waiting room my consulting hours often get delayed which is a nuisance for my patients. I have found a solution for this problem using my VR computer software.
Mr. Claassen is an elderly man with incontinence and urge problems. The delay makes him nervous and stimulates his urge even more.
He's afraid to visit the toilet because he thinks he will be bypassed by other patients. I chose him for my experiment with time for this reason. I programmed his watch through the VR software to tick just some slower than in reality. All movements in his VR view are happening slower than in reality. It doesn't look unnatural because his brain is misled. In this way I could slow

down time for him with 50% and he's experiencing not a delay of 30 minutes but of 15 minutes. In this way I helped him out of his stress. Later I expanded my experiment to a larger group of patients, which made me control my office time better without frustrating my patients. They will be shocked standing outside, reading their watches and discovering how much real time has passed. I can also play with space. I can put fake virtual patients on certain chairs

1

keeping them free and delete certain events (like the throwing up of a patient or a crying baby).

With this thought experiment I want to show you what can be simulated by a computer and what not. So, the ego of a person can never be simulated by a computer, but events, time and space can.

These experiences happen to be depending on the sensory part of our brain, let's say the "machine "part. Thoughts and the feeling to be an existing person can't, because the computer isn't conscious and though the programmer is, he can't be someone else and doesn't know how to program that "soul" part in a computer.

Chapter twenty-two

Why time machines are impossible

Albert Einstein taught us why time machines can't exist and his reasoning was surprisingly not put as a physicist, but purely on philosophical grounds in the physicist's and mathematician's perspective time would flow from past to future and backwards. But the second thermodynamic law forbids the last possibility. The law of entropy states that entropy is always increasing, otherwise said, the state of disorder in the universe is increasing until a new equilibrium is found. If you throw the content of a cup of water on your kitchen floor the splashing waterdrops will never ever find their way back exactly where they came from, into the cup of water. The parts of a Boeing 747 will never find their way back into the place it was constructed in the factories all over the world. That's true on our macrocosmic scale and very obvious, but not on the microcosmic quantum scale. Quanta can go back in time (a very tiny bit). Time crystals which maintain time stored, are going back and forth in time and can really be made. But that's just laboratory stuff and doesn't bring us to building a time machine.

So, what was Einstein's argument against it?
Firstly, time is an illusion created by consciousness anyway. It's just a certain order in which events happen in the context of conscious experience and not an objective physical phenomenon. For physicists and mathematicians, it's no problem to delete all time units from all formulas, but it wouldn't be practical building Boeing 747 's.
Time is not constant, because if you move faster, time goes slower and also when you're in a strong gravity field. Time stops when you reach the speed of light and when you reach the event horizon of a black hole.
There is no universal clock like Greenwich mean time.
So, it might be possible to reach almost the speed of light and hardly getting older and travel 1000 lightyears and experiencing a week's travel! Theoretically yes, but technically impossible, because the amount of energy needed for a spaceship getting more mass when speed accelerates (special relativity) is not producible.
Even if it was, you could be possibly visiting some planets and an intelligent species over there and get their knowledge, return to earth 2000 years later, but did you experience a time machine then? No, because you only made a 2000-year trip not getting anything older.
Space travel doesn't provide time machines.

Suppose time travel was possible, then you would be confronted by famous paradoxes.

1

1 *The grandfather paradox.*
By killing your grandfather in the past, you would prevent your own birth and so prevent the killing of your grandfather.

2. *The butterfly paradox.* If you would kill a butterfly in the age of the dinosaurs this would have impact on the long run on the history of our planet (chaos theory) Any tiny trivial intervention would have that effect, let alone walking around there and changing a lot of things. You would make a mess of our present time.

3 *The funeral paradox.* Going to the future would make you able to be on your own funeral, but how can you be dead and alive in the same time?

All these arguments have to do with the nature of consciousness. Your consciousness is undividable, unique in time, place and history and cannot be split up in a " time travel ego " nor a " future" or " past "ego. It also shows the past is unchangeably unique for the universe and certainly cannot be changed by a time traveller. The other question remains.

Can the future be changed? Well, we're doing that every minute of the day, don't we? Are we totally in control though? Many events happen being not under control of our free will (if this exists at all).

Many tiny trivial events and great events change the future all the time, from killing a butterfly to starting a world war.

Do they also have a unique timing and a unique significance? There are a lot of arguments for that in this perspective. The exact time you die and you're born, already influences the behaviour of the undertaker and the midwife and so in the long run millions of things in consequence. Even these events can't be left to the arbitrariness of blunt coincidence if you ask me. So probably the future already exists, but we are just not on that point of experience yet, because we're trapped in time.

Until now we have had no visitors from the future nor past. Together with the arguments mentioned, time travel is forbidden by the laws of nature. What would happen if you step in a theoretical time machine? I'm sure you would die if you pass the time of your birth or death.

But what would happen if you visit a time spot somewhere in your life span? I'm sure you will never have known about your time machine the moment you step out of it. You will suddenly be just who you are at that point of time in your life, for example the toddler, schoolchild, old man which is you and nothing but you and you will never observe or know about some time machine.

This sounds magical, but is less magical then seeing yourself split up in a time machine "me" and a real "me". This is exactly what Einstein thought about time machines.

Our timeline of events is unchangeable for ever. The story was written in an instant without the use of any time, but with the existence of multiple universes in the superposition state of the quantum world.

Our universe popped out in an instant, completely written and perfectly finished, but just there just to be experienced by conscious observers, like us.

Desert

In this book many arguments against the reductionistic method of science are brought together and I'm aiming for a more holistic future view of science. Holism holds a vague purport in it, but actually that's a misunderstanding. The only way I can support this argument is with a formula:

$$í \times \pi = \omega$$

Which means I (upsilon) x P (pie) = O (omega)
Or: Information(I) multiplied with process(P) makes outcome (O).

In words, this means that a piece of **Information**, which is **processed** somehow, has an **outcome** which is more complex than the original piece of information.

Just as Stuart Kaufman pointed out in his book "The law of self- organization", I agree with him that simple information is worked up into more complex information by "some process" everywhere in the universe.
Quanta making sub elementary particles, making atoms, making molecules, making cells, making tissues, making organs, making bodies , making brains, making minds, making….God ?
So, our brain being the most complex structure we're familiar with, brings us in an even more complex state, transcending the so called physical border, making the human mind operating in an immaterial world. As I pointed out several times, this view of having a material and an immaterial world is wrong and the grand misunderstanding in physics. There is no physical border, because there's no physical world! There's only an immaterial world with many layers of complexity in one universal mind. This might be the universal truth and í x π = ω the most elementary law.

Mattees van Dijk

20--2-2019

1